# EMERGING THERMAL AND NONTHERMAL TECHNOLOGIES IN FOOD PROCESSING

# EMERGING THERMAL AND NONTHERMAL TECHNOLOGIES IN FOOD PROCESSING

*Edited by*

**Prem Prakash Srivastav, PhD**
**Deepak Kumar Verma, PhD**
**Ami R. Patel, PhD**
**Asaad Rehman Al-Hilphy, PhD**

APPLE ACADEMIC PRESS

Apple Academic Press Inc.
4164 Lakeshore Road
Burlington ON L7L 1A4
Canada

Apple Academic Press, Inc.
1265 Goldenrod Circle NE
Palm Bay, Florida 32905
USA

**Library and Archives Canada Cataloguing in Publication**

Title: Emerging thermal and nonthermal technologies in food processing / edited by Prem Prakash Srivastav, PhD, Deepak Kumar Verma, PhD, Ami R. Patel, PhD, Asaad Rehman Al-Hilphy, PhD.
Names: Srivastav, Prem Prakash, editor. | Verma, Deepak Kumar, 1986- editor. | Patel, Ami R., editor. | Al-Hilphy, Asaad Rehman, editor.
Description: Includes bibliographical references and index.
Identifiers: Canadiana (print) 20200175262 | Canadiana (ebook) 20200175343 | ISBN 9781771888318 (hardcover) | ISBN 9780429297335 (ebook)
Subjects: LCSH: Food industry and trade—Technological innovations.
Classification: LCC TP371.2 .E44 2020 | DDC 664/.028—dc23

CIP data on file with US Library of Congress

# About the Editors

**Prem Prakash Srivastav, PhD**
*Associate Professor, Department of
Agricultural and Food Engineering Department,
Indian Institute of Technology Kharagpur,
West Bengal, India*

Prem Prakash Srivastav, PhD, is Associate Professor of Food Science and Technology in the Agricultural and Food Engineering Department at the Indian Institute of Technology Kharagpur (West Bengal), India. His research interests include development of specially designed convenience, functional, and therapeutic foods; extraction of nutraceuticals; and development of various low-cost food-processing machineries. He has organized many sponsored short-term courses and completed sponsored research projects and consultancies. He has published various research papers in peer-reviewed international and national journals and proceedings and many technical bulletins and monographs as well. Other publications include books and book chapters along with many patents. He has attended, chaired, and presented various papers at international and national conferences and delivered many invited lectures at various summer/winter schools. Dr. Srivastav has received best poster paper awards at various professional conferences and a is life member of various professional bodies, including the International Society for Technology in Education, Association of Food Scientists & Technologists India, Indian Dietetic Association, Association of Microbiology *India*, American Society of Agricultural and Biological Engineers, and the Institute of Food Technologists (USA). He graduated from Gorakhpur University, Gorakhpur, India, and received his MSc degree with major in Food Technology and a minor in Process Engineering from G. B. Pant University of Agriculture and Technology, Pantnagar, India. He was awarded a PhD from the Indian Institute of Technology Kharagpur. He teaches various undergraduate-, postgraduate-, and PhD-level courses and has guided many student research projects.

## Deepak Kumar Verma, PhD

*Research Scholar, Department of Agricultural and Food Engineering, Indian Institute of Technology, West Bengal, India*

Deepak Kumar Verma is an agricultural science professional and is currently a PhD Research Scholar in the specialization of food processing engineering in the Agricultural and Food Engineering Department, Indian Institute of Technology, Kharagpur (WB), India. In 2012, he received a DST-INSPIRE Fellowship for PhD study by the Department of Science & Technology (DST), Ministry of Science and Technology, Government of India. Mr. Verma is currently working on the research project "Isolation and Characterization of Aroma Volatile and Flavoring Compounds from Aromatic and Non-Aromatic Rice Cultivars of India." His previous research work included "Physico-Chemical and Cooking Characteristics of Azad Basmati (CSAR 839-3): A Newly Evolved Variety of Basmati Rice (*Oryza sativa* L.)". He earned his BSc degree in agricultural science from the Faculty of Agriculture at Gorakhpur University, Gorakhpur, and his MSc (Agriculture) in Agricultural Biochemistry in 2011. Apart from his area of specialization in plant biochemistry, he has also built a sound background in plant physiology, microbiology, plant pathology, genetics and plant breeding, plant biotechnology and genetic engineering, seed science and technology, food science and technology etc. In addition, he is a member of different professional bodies, and his activities and accomplishments include conferences, seminar, workshop, training, and also the publication of research articles, books, and book chapters.

### Ami R. Patel, PhD

*Assistant Professor, Division of Dairy and Food Microbiology, Mansinhbhai Institute of Dairy & Food Technology, Dudhsagar Dairy Campus, Gujarat, India*

Ami R. Patel, PhD, is an Assistant Professor in the Division of Dairy and Food Microbiology at the Mansinhbhai Institute of Dairy & Food Technology, Dudhsagar Dairy Campus, Gujarat, India. Her expertise is in a specialized area that involves isolation, screening, and characterization of exopolysaccharides from potential probiotic cultures and employing them for food and health applications. In addition, she is engaged with research and teaching undergraduate and postgraduate students. She has authored a number of peer-reviewed papers and technical articles in international and national journals as well as book chapters, proceedings, and technical bulletins. She has received a number of awards and honors, including the Vice Chancellor Gold Medal for her PhD work; the BiovisinNxt11 fellowship to attend an international conference at Lyon, France; and an Erasmus Mundus scholarship for three years to visit Lund University, Sweden, as guest researcher. She is serving as an expert reviewer for several scientific journals and is a member of academic and professional organizations, including the Indian Dairy Association and the Swedish South Asian Studies Network (fermented foods). She earned her BSc (Microbiology) and her MSc (Microbiology) degrees from Sardar Patel University, Vallabh Vidyanagar, Gujarat, and her PhD in Dairy Microbiology from the Dairy Department of SMC College of Dairy Science, Anand Agricultural University, Gujarat, India.

## Asaad Rehman Al-Hilphy, PhD

*Professor, Department of Food Science, College of Agriculture, University of Basrah, Basra, Iraq*

Asaad Rehman Al-Hilphy, PhD, is a Professor in the Department of Food Science at the College of Agriculture at the University of Basrah, Iraq. His research interests include development of food processing equipment, solar food drying, extraction of essential oils, thermal and nonthermal food processing engineering, and water desalination. He has organized many sponsored short-term courses and completed sponsored research projects and consultancies. He has published various research papers in peer-reviewed international and national journals and proceedings and many technical bulletins and monographs as well. Other publications include books and book chapters along with many patents. He has attended, chaired, and presented various papers at international and national conferences and delivered many invited lectures at various summer/winter schools. He is a life member of tje International Society of Food Engineering (USA) and the International Society for Development and Sustainable (Japan). He graduated from the Agricultural Machinery Department of Basrah University, Iraq, where he received his MSc degree with a major in Agronomy and a minor in Design of Tillage equipment from the Agriculture College. He also was awarded his PhD from the University of Basrah. He teaches various undergraduate-, postgraduate-, and PhD-level courses and has guided many student research projects.

# Contents

Dedication .................................................................................................... *xi*

Contributors ............................................................................................... *xiii*

Abbreviations ............................................................................................ *xvii*

Symbols .................................................................................................... *xxi*

Preface .................................................................................................... *xxiii*

1. **Evaluation of Thermal Processing in the Dairy Industry Using Milk Enzymes** ........................................................................................... 1

   Loredana Dumitraşcu, Gabriela Râpeanu, and Nicoleta Stănciuc

2. **Microwave Heating: Alternative Thermal Process Technology for Food Application** ........................................................................... 25

   Deepak Kumar Verma, Naveen Kumar Mahanti, Mamta Thakur, Subir Kumar Chakraborty, and Prem Prakash Srivastav

3. **Effects of Drying Technology on Physiochemical and Nutritional Quality of Fruits and Vegetables** ......................................... 69

   Deepak Kumar Verma, Mamta Thakur, Prem Prakash Srivastav, Vahid Mohammadpour Karizaki, and Hafiz Ansar Rasul Suleria

4. **Disinfection of Drinking Water by Low Electric Field** ...................... 117

   Asaad Rehman Saeed Al-Hilphy, Nawfal A. Alhelfi, and Saher Sabih George

5. **Removal Cholesterol from Minced Meat Using Supercritical $CO_2$** .... 131

   Asaad Rehman Saeed Al-Hilphy, Munir Abood Jasim Al-Tai, and Hassan Hadi Mehdi Al Rubaiy

6. **Microwave-Convective Drying of Ultrasound Osmotically Dehydrated Tomatoes** ........................................................................... 157

   João Renato de Jesus Junqueira, Francemir José Lopes, Jefferson Luiz Gomes Corrêa, Kamilla Soares de Mendonça, Randal Costa Ribeiro, and Bruno Elyeser Fonseca

7. **Ultrasound-Assisted Osmotic Dehydration in Food Processing: A Review** ................................................................................................. 175

   Vahid Mohammadpour Karizaki

8.  Hydrodynamic Cavitation Technology for Food Processing
    and Preservation ...................................................................................... 199

    Naveen Kumar Mahanti, Subir Kumar Chakraborty, S. Shiva Shankar, and
    Ajay Yadav

9.  High Pressure Processing (HPP): Fundamental Concepts,
    Emerging Scope, and Food Application............................................... 225

    Deepak Kumar Verma, Mamta Thakur, Jayant Kumar, Prem Prakash Srivastav,
    Asaad Rehman Saeed Al-Hilphy, Ami R. Patel, and Hafiz Ansar Rasul Suleria

10. Induced Electric Field (IEF) as an Emerging Nonthermal
    Techniques for the Food Processing Industries:
    Fundamental Concepts and Application............................................... 259

    Na Yang and Dan-Dan Li

Index................................................................................................................... 279

# Dedication

---

**This book is dedicated to the memory of remarkable people,**

**My Guru**

*Professor* **H. Das**

*AgFE Department, IIT Kharagpur, West Bengal, India*

# Contributors

**Nawfal A. Alhelfi**
Department of Food Science, College of Agriculture, University of Basrah, Basra City, Iraq, Mobile: +00-96-47703109177, E-mail: nawfalalhelfi@gmail.com

**Asaad Rehman Saeed Al-Hilphy**
Department of Food Science, College of Agriculture, University of Basrah, Basra City, Iraq, Mobile: +00-96-47702696458, E-mail: aalhilphy@yahoo.co.uk

**Hassan Hadi Mehdi Al Rubaiy**
Department of Food Science, College of Agriculture, University of Basrah, Basrah City, Iraq, Mobile: +00-96-477035594409, E-mail: drhassanhadi78@gmail.com

**Munir Abood Jasim Al-Tai**
Department of Food Science, College of Agriculture, University of Basrah, Basrah City, Iraq, Mobile: +00-96-47713188406, E-mail: dr.munir2000@yahoo.com

**Subir Kumar Chakraborty**
Agro Produce Processing Division, ICAR-Central Institute of Agricultural Engineering, Bhopal – 462038, Madhya Pradesh, India, Mobile: +91-7552521209, E-mail: Subir.Kumar@icar.gov.in

**Jefferson Luiz Gomes Corrêa**
Food Science Department, Federal University of Lavras, Lavras, Minas Gerais State, Brazil, Tel.: +55-3538291393, Fax: +55-3538291391, E-mail: jefferson@dca.ufla.br

**Loredana Dumitraşcu**
Faculty of Food Science and Engineering, "Dunărea de Jos" University of Galati, Domnească Street – 111, 800201, Galati, Romania, Tel.: +40-0336130183, Mobile: +40-0758645858, Fax: +40-0236-460165, E-mail: loredana.dumitrascu@ugal.ro

**Bruno Elyeser Fonseca**
Engineering Department, Federal University of Lavras, Lavras, Minas Gerais State, Brazil, Tel.: +55-3538291393, Fax: +55-3538291391, E-mail: brunoelyezerfonseca@yahoo.com.br

**Saher Sabih George**
Department of Food Science, College of Agriculture, University of Basrah, Basra City, Iraq, Mobile: +00-96-47705652900, E-mail: Saher_sg@yahoo.com

**João Renato de Jesus Junqueira**
Food Science Department, Federal University of Lavras, Lavras, Minas Gerais State, Brazil, Tel.: +55-3538291393, Fax: +55-3538291391, E-mail: jrenatojesus@hotmail.com

**Vahid Mohammadpour Karizaki**
Chemical Engineering Department, Quchan University of Advanced Technology, Quchan, Iran, Mobile: 00989477167335, Tel.: 00989354808629, E-mail: mohammadpour_vahid@yahoo.com

**Jayant Kumar**
Agricultural and Food Engineering Department, Indian Institute of Technology,
Kharagpur – 721302, West Bengal, India, Mobile: +91-7878750780,
E-mail: Jayant.iitkgpian@gmail.com

**Dan-Dan Li**
School of Food Science and Technology, Jiangnan University, Wuxi – 214122, Jiangsu Province,
China, Tel./Fax: +01186-510-85919182, E-mail: lidandanthora@gmail.com

**Francemir José Lopes**
Food Science Department, Federal University of Lavras, Lavras, Minas Gerais State, Brazil,
Tel.: +55-3538291393, Fax: +55-3538291391, E-mail: francemirlopes@yahoo.com.br

**Naveen Kumar Mahanti**
Agricultural Processing and Structures Division, Indian Agricultural Research Institute,
New Delhi – 110012, India, Mobile: +91-8500898426, E-mail: naveeniitkgp13@gmail.com

**Kamilla Soares de Mendonça**
Food Science Department, Federal University of Lavras, Lavras, Minas Gerais State, Brazil,
Tel.: +55-3538291393, Fax: +55-3538291391, E-mail: keamendonca@msn.com

**Ami R. Patel**
Division of Dairy and Food Microbiology, Mansinhbhai Institute of Dairy and Food Technology-
MIDFT, Dudhsagar Dairy Campus, Mehsana – 384 002, Gujarat, India,
Tel.: +00-91-2762243777 (O), Mobile: +00-91-9825067311, Fax: +91-02762-253422,
E-mail: amiamipatel@yahoo.co.in

**Gabriela Râpeanu**
Faculty of Food Science and Engineering, "Dunărea de Jos" University of Galati,
Domnească Street – 111, 800201, Galati, Romania, Tel.: +40-0336130183,
Mobile: +40-0742038288, Fax: +40-0236-460165, E-mail: gabriela.rapeanu@ugal.ro

**Randal Costa Ribeiro**
Engineering Department, Federal University of Lavras, Lavras, Minas Gerais State, Brazil,
Tel.: +55-3538291393, Fax: +55-3538291391, E-mail: randalribeiro8@gmail.com

**S. Shiva Shankar**
Department of Post-Harvest and Food Engineering, G. B. Pant University of Agriculture and
Technology, Udham Singh Nagar, Pantnagar, Uttarakhand – 263153, India,
Mobile: +91-9550579386, E-mail: shiva14cae@gmail.com

**Prem Prakash Srivastav**
Agricultural and Food Engineering Department, Indian Institute of Technology Kharagpur – 721302,
West Bengal, India, Tel.: +91-3222281673, Fax: +91-3222282224, E-mail: pps@agfe.iitkgp.ernet.in

**Nicoleta Stănciuc**
Faculty of Food Science and Engineering, "Dunărea de Jos" University of Galati,
Domnească Street – 111, 800201, Galati, Romania, Tel.: +40-0336130183,
Mobile: +40-0729270954, Fax: +40-0236-460165, E-mail: nicoleta.stanciuc@ugal.ro

**Hafiz Ansar Rasul Suleria**
UQ Diamantina Institute, Translational Research Institute, Faculty of Medicine,
The University of Queensland, 37 Kent Street Woolloongabba, Brisbane, QLD 4102,
Australia, Mobile: +61-470-439-670, E-mail: hafiz.suleria@uqconnect.edu.au

**Mamta Thakur**
Department of Food Engineering and Technology, Sant Longowal Institute of Engineering and Technology, Longowal – 148106, Punjab, India, Tel.: +91-8219831376; 8352895496, Fax: +91-1672-280057, E-mails: thakurmamtafoodtech@gmail.com; mamta.ft@gmail.com

**Deepak Kumar Verma**
Agricultural and Food Engineering Department, Indian Institute of Technology, Kharagpur – 721 302, West Bengal, India, Tel.: +91-3222-281673, Mobile: +00-91-7407170260, Fax: +91-3222-282224, E-mails: deepak.verma@agfe.iitkgp.ernet.in; rajadkv@rediffmail.com

**Ajay Yadav**
Scientist, Center of Excellence for Soybean Processing and Utilization, ICAR-Central Institute of Agricultural Engineering, Bhopal – 462038, Madhya Pradesh, India, Mobile: +91-8818805007, E-mail: ajyadav007@gmail.com

**Na Yang**
School of Food Science and Technology, Jiangnan University, Wuxi – 214122, Jiangsu Province, China, Tel./Fax: +01186-510-85919182, E-mail: yangna@jiangnan.edu.cn

# Abbreviations

| | |
|---|---|
| AA | antioxidant activity |
| AC | alternative current |
| AD | air drying |
| ALP | alkaline phosphatase |
| AM | air-microwave |
| AP | apple pomace |
| BBB | broccoli-based bars |
| CD | convective drying |
| CF | centrifugal force |
| CFU | colony-forming unit |
| CLELAB | CIE $l^*$ $a^*$ $b^*$ |
| ClO$_2$ | dioxide chlorine |
| CPD | convective pre-drying |
| DAA | dehydroascorbic acid |
| DC | direct current |
| DNA | deoxyribonucleic acid |
| DPPH | 2,2-diphenyl-1-picrylhydrazyl |
| *E. coli* | *Escherichia coli* |
| EC | electrical conductivity |
| EMC | equilibrium moisture content |
| EMF | electromotive force |
| EU | European Union |
| FAO | Food and Agriculture Organization |
| FD | freeze-drying |
| FRAP | ferric reducing antioxidant power |
| GLS | glucosinolates |
| GSY | Greek-style yogurt |
| GT | glutamyltransferase |
| HA | hot air |
| HACD | hot air convective drying |
| HAD | hot air drying |
| HD | hybrid drying |
| HDC | hydrodynamic cavitation |

| | |
|---|---|
| HHAIB | humidity hot air impingement blanching |
| HHP | high hydrostatic processing |
| HIPEF | high-intensity pulsed electric field |
| HPH | high pressure homogenizer |
| HPP | high pressure processing |
| HS | hypertonic solution |
| HTD | hydro thermodynamic |
| HTST | high temperature short time |
| HW | hot water |
| HWB | hot water blanching |
| IDF | International Dairy Federation |
| IEF | induced electric field |
| IRB | infrared blanching |
| ISI | innovative steam injection |
| IVSD | *in-vitro* starch digestibility |
| LP | lactoperoxidase |
| MAP | microwave assisted pasteurization |
| MC | moisture content |
| MCWC | microwave circulated water combination |
| MEF | moderate electric field |
| MMF | magnetomotive force |
| MW | microwave |
| MWAD | microwave-assisted air drying |
| MWB | microwave blanching |
| MWD | microwave drying |
| MWFD | microwave-assisted freeze-drying |
| MWSB | microwave-assisted spouted bed |
| MWVD | microwave-assisted vacuum drying |
| NaCl | sodium chloride |
| OD | osmotic dehydration |
| OH | ohmic heating |
| ORAC | oxygen radical absorbance capacity |
| OU | oil uptake |
| PCs | phenolic compounds |
| PEF | pulse electric fields |
| PID | proportional integral derivative |
| PL | pulsed light |
| PLC | programmable logic controller |

| PME | pectin methylesterase |
|---|---|
| POD | peroxidase |
| PPO | polyphenol oxidase |
| PR | pulse rate |
| PSPs | purple sweet potatoes |
| QP | quinolyl phosphate |
| RH | relative humidity |
| RMS | root mean square |
| RMSE | root mean square error |
| RO | reverse osmosis |
| RR | rehydration ratio |
| RS | resistant starch |
| RSM | response surface methodology |
| SB | steam blanching |
| SC-CO$_2$ | supercritical CO$_2$ |
| SE | standard error |
| SF | sorghum flour |
| SG | solid gain |
| SL | shelf-life |
| SPSS | statistical package for social science |
| SS | soluble starch |
| TDF | total dietary fiber |
| TEAC | Trolox equivalent antioxidant capacity |
| TI | time-temperature indicator |
| TPC | total phenolic compounds |
| TPMC | carbon dioxide treated milk protein concentrate |
| TSB | tryptic soy broth |
| TSC | total starch content |
| UAOD | ultrasound-assisted osmotic dehydration |
| UHP | ultra-high pressure |
| UHT | ultra-high temperature |
| USA | United States of America |
| USDA | United States Department of Agriculture |
| USFDA | United States Food and Drug Administration |
| USNAC | United State National Advisory Committee |
| UV | ultraviolet |
| UVR | ultra-violet radiation |
| UW | ultrasonic waves |

| VFCs | volatile flavor compounds |
| VMD | vacuum-microwave-dried |
| WL | water loss |
| WR | weight reduction |

# Symbols

| | |
|---|---|
| μm | micrometer |
| μs | microsecond |
| $A/A_0$ | ratio between residual enzymatic activity and initial enzymatic activity |
| $a_w$ | water activity |
| cm | centimeter |
| $C_v$ | cavitation number |
| d.b. | dry basis |
| $D_{eff}$ | effective diffusivity |
| $E_a$ | activation energy |
| $e_p$ | induced electromotive force |
| $E_p$ | induced voltage in the primary coil |
| $E_S$ | induced voltage in the secondary coil |
| $E_{S1}$ | induced voltage at the secondary coil 1 |
| $E_{S2}$ | induced voltage at the secondary coil 2 |
| eV | electron volt |
| $f$ | frequency |
| $i_0$ | exciting current |
| $i_m$ | magnetizing current |
| J | joule |
| k | drying rate |
| kGy | kiloGrays |
| kHz | kiloHertz |
| $K_m$ | Michaelis-Menten constant |
| kV | kiloVolt |
| L | characteristic length |
| log | logarithm |
| min | minute |
| Mpa | mega Pascal |
| $M_r$ | dimensional moisture content |
| n | adjustment parameter |
| nm | nanometer |
| $N_p$ | primary coil turns |

| | |
|---|---|
| $N_{P1}$ | number of turns in the primary coil 1 |
| $N_{P2}$ | number of turns in the primary coil 2 |
| $N_s$ | secondary coil turns |
| $r^2$ | correlation coefficient |
| t | time |
| U/mL | units per milliliter |
| $U_{ab}$ | voltage between two equipotential cells of liquid sample |
| $U_{Coil}$ | divided voltage on the secondary coil of liquid sample |
| $U_S$ | output voltage |
| $U_{S1}$ | terminal voltage at the secondary coils 1 |
| $U_{S2}$ | terminal voltages at the secondary coils 2 |
| V | volt |
| $V_a$ | transient potential at terminal $a$ of the secondary coil |
| $V_b$ | transient potential at terminal $b$ of the secondary coil |
| $v_p$ | instantaneous value of the applied voltage |
| X | moisture content |
| $Z_{Coil}$ | internal impedance of the coil |
| $Z_{Voltmeter}$ | external impedance of the load resistor |
| $\phi_m$ | mutual flux |
| $\phi_{mp}$ | peak value of the mutual flux |

# Preface

The processing of foods chiefly involves a few major activities, such as mincing/macerating, liquefaction, emulsification, and cooking (such as boiling, frying, broiling, roasting, or grilling); pickling, heating (such as pasteurization), and canning plus packaging. Food processing is the key stage that principally affects the physical or biochemical properties of foods in conjunction with determining the shelf-life and safety of the finished product. The outcome of recent innovations in thermal as well as nonthermal technologies that are specifically applied for potable water and fluid foods (milk, juice, soups, etc.), well documented for their high bioavailability of macro- and micronutrients, are very promising.

The technologies of processing of wholesome foods require considerable resources and expertise. Many diverse technologies are employed during the processing of different foods. With the changing consumption patterns, nutrition awareness, product popularity, and enhanced shelf-life, newer food process technologies are also being evolved. In recent years, novel thermal and nonthermal technologies have been developed. Most nonthermal technologies are developed as an alternative to thermal processing, while still meeting required safety or shelf-life demands and minimizing the effects on its nutritional and quality attributes.

This book, *Emerging Thermal and Nonthermal Technologies in Food Processing*, provides a comprehensive overview of thermal and nonthermal processing of foods with innovations and new technologies. There is a total of 10 different chapters in this book, in which Chapter 1 covers the inactivation kinetics of alkaline phosphatase, γ-glutamyltransferase, and lactoperoxidase during the thermal processing of milk where milk fat is associated with the enzymatic activity. The alkaline phosphatase, together with γ-glutamyltransferase, are considered as markers for milk pasteurization, while lactoperoxidase seems to be a good indicator for marking the region of intermediate pasteurization.

Chapter 2 includes the application of microwaves in heating, drying, pasteurization, sterilization, blanching, baking, cooking, thawing, and microwave-assisted extraction of compounds in order to overcome the several constraints related to conventional heating.

Chapter 3 deals with the physical, chemical, nutritional, sensorial, and rehydration characteristics of food products as influenced by drying. This chapter also reviews the need for novel drying techniques to maintain food quality and safety.

Chapter 4 investigates the potential of the low-electric field in the disinfection of drinking water and reports the significant reduction in total bacteria and total coliform bacteria count, *E. coli,* and survival microorganisms to the total ratio on increasing the electrical field strength.

Chapter 5 discusses the detailed mechanism of cholesterol exclusion from minced meat by altering the temperature and pressure of supercritical carbon dioxide.

Chapter 6 provides the knowledge about the microwave-convective drying of tomato slices treated with ultrasound-assisted osmotic dehydration to maintain the quality, especially the color and lycopene content.

Chapter 7 highlights the role of ultrasound in osmotic dehydration of food products to improve the process efficiency and quality.

Chapter 8 focuses on the potential of hydrodynamic cavitation in food and water processing as an alternative to the ultrasonic cavitation technique. However, this novel technology, in addition to being highly energy-efficient, makes the contacting parts to erode which however, can be controlled by concentrating the cavitation towards the Center of the stream.

Chapter 9 presents the technology, principle, and scope of high-pressure processing in the preservation, safety, and quality improvement of food products. The content also involves the commercial applications and current research prevailing around the globe on the high-pressure processing, either single or in combination with other processing methods like ultrasound, gamma-irradiation, and heat.

Chapter 10 describes the technical aspects, principles, processing parameters, and potential applications of the induced electric field in the food processing and preservation.

Thus, this volume, *Emerging Thermal and Nonthermal Technologies in Food Processing*, brings together information on fluid and microbial and quality dynamics for fluid foods to facilitate process validation and technology adoption. Worldwide adoption of these novel technologies will benefit consumers in terms of enhanced food safety labels, nutritional security, and value-added products at a reasonable cost.

We are sure that this book will serve as a comprehensive reference and useful guide for students, educators, researchers, food processors, and industry personnel looking for up-to-date insight into the field. Additionally, the covered range of techniques for by-product utilization will provide engineers and scientists working in the food industry with a valuable resource for their work.

It is hoped that this edition will stimulate discussions and generate helpful comments to improve upon future editions.

Finally, with great pleasure, I would like to extend my sincere thanks to all the eminent renowned scientists, researchers, and professors who have contributed through their magnificent and tireless research with dedication, persistence, and cooperation, which enabled us to present this work with detail and accuracy of information in the form, which has made our task easy as editors a pleasure.

We acknowledge Almighty God, who provided all the inspirations, insights, positive thoughts, and channels to complete this book project.

**—Prem Prakash Srivastav, PhD**
**Deepak Kumar Verma, PhD**
**Ami R. Patel, PhD**
**Asaad Rehman Al-Hilphy, PhD**

# CHAPTER 1

# Evaluation of Thermal Processing in the Dairy Industry Using Milk Enzymes

LOREDANA DUMITRAŞCU, GABRIELA RÂPEANU, and
NICOLETA STĂNCIUC

*Faculty of Food Science and Engineering,*
*"Dunărea de Jos" University of Galati, Galati, Romania,*
*E-mail: loredana.dumitrascu@ugal.ro (L. Dumitraşcu);*
*E-mail: gabriela.rapeanu@ugal.ro (G. Râpeanu);*
*E-mail: nicoleta.stanciuc@ugal.ro (N. Stănciuc)*

## 1.1 INTRODUCTION

Milk is one of the most complex food that contains mainly water, proteins, fats, carbohydrates, minerals, vitamins, and enzymes (Niamah and Verma, 2017; Verma et al., 2017). For centuries, milk is considered one of the most valuable natural food from the human dict (Al-Hilphy ct al., 2016). On the other hand, milk nutrients, neutral pH and high water activity provide excellent conditions for many pathogenic microorganisms to be developed, whose multiplication is dependent on temperature and as well as on competing microorganisms (Claeys et al., 2014). Therefore, heat treatment should be applied to guarantee its microbial safety and stability. Moreover, the functional and nutritional properties of milk are changed during heating as a result of various competitive and interdependent reactions that are dependent on heating conditions, milk composition, and origin. The main heat treatments applied in dairy industry include: thermalization (57–68°C, 15–20 s), HTST pasteurization (71–74°C, 15–40 s), sterilization (110–120°C, 20 min.), indirect UHT (135–140°C, 6–10 s), direct UHT (140–150°C, 2–4 s) and ISI (innovative steam injection, 150–200°C, < 0.1 s).

A comprehensive review was reported by Claeys et al. (2014), where the authors presented the risks and benefits associated with the consumption of raw and processed cow milk, considering microbiological and nutritional properties. Most of the dairy products available for consumption are obtained from cow milk. The increased interest of consumers for healthy diets generated increased attention for milk and dairy products of nonbovine origin, and especially for caprine and ovine origin. Caprine milk possesses stronger antimicrobial, immunological, and antibacterial system and higher digestibility compared with ovine or bovine milk (Slacanac et al., 2010). The growing interest in goat milk as an alternative food for infants with food allergies should be supported by appropriate studies showing its suitability for human consumption and in terms of milk safety. In most countries, assessment of the milk of non-bovine origin is not yet introduced in routine testing programs. Regulation (EC) 853/2004 lays down regulatory microbial criteria for total plate count and somatic cells, as well as health and hygienic requirements for animal production and production facilities, respectively. In regard to heat treatments applied in the dairy industry, European regulations are mainly based on microbiological platforms, which are time-consuming and expensive to control. Fast, easy control instruments to assess heating process efficiency and severity are scarce.

In the entitled chapter, research findings about some milk enzymes from the bovine and nonbovine origin are presented. Information regarding the activity of alkaline phosphatase (ALP), $\gamma$-glutamyltransferase (GT) and lactoperoxidase (LP) in bovine, caprine, and ovine milk, as well as the effect of milk species on the thermal inactivation of these enzymes, is briefly discussed.

## 1.2   TIME TEMPERATURE INDICATORS

Milk enzymes are distributed in different milk phases (Table 1.1), many of them possessing technological implications (Fox, 2006) as following:

1. Alteration (lipase, acid phosphatase, xanthine oxidase) or preservation (LP, sulphydryl oxidase, superoxide dismutase) of milk quality;
2. Indicators for the assessment of milk thermal processing (ALP, $\gamma$-GT, and LP);

3. Mastitis indicators (catalase, acid phosphatase, β-*N*-acetylglucosaminidase), whose concentration increase during mastitis infection;
4. Antimicrobial activity (lysozyme, LP);
5. Commercial source of enzymes (ribonuclease and LP).

**TABLE 1.1**  Indigenous Enzymes in Cow Milk

| Enzyme | EC | Source | Distribution in Milk Phases |
|---|---|---|---|
| Plasmin | 3.4.21.7 | Blood | Mainly in casein micelles |
| *Lysosomal Proteinases* | | | |
| Cathepsin D | 3.4.23.5 | Somatic cells | Acid whey |
| Cathepsin B | 3.4.22.1 | | |
| Other proteinases | - | | |
| Lipoprotein lipase | 3.1.1.34 | Mammary gland | Casein micelles |
| *Phospohydrolases* | | | |
| Alkaline phosphatase | 3.1.3.1 | Mammary gland | Milk fat globule membrane / skim milk |
| Acid phosphatase | 3.1.3.2 | | |
| Ribonuclease | 3.1.27.5 | Blood | Serum |
| *Oxidases* | | | |
| Lactoperoxidase | 1.11.1.7 | Mammary gland | Serum |
| Catalase | 1.11.1.6 | Somatic cells | Cream/skim milk |
| Xanthin oxidase | 1.1.3.22 | Blood | Milk fat globule membrane |
| Superoxide dismutase | 1.15.1.1 | | Serum |
| γ-glutamyltransferase | 2.3.2.2 | Mammary gland | Membrane material in skim milk |
| N-acetyl-β-glucosaminidase | 3.2.1.52 | Somatic cells | Skim milk |
| Amylases (diastase, α-amylase) | 3.2.1.1 | Blood | Serum/Skim milk |
| Lysozyme | 3.2.1.17 | Lysosomal enzymes | Serum |

(*Source*: Reprint with permission from Moatsou, 2010. © John Wiley and Sons.)

The evaluation of heat treatment efficiency can be carried out using three different approaches:

1. ***In-Situ* Evaluation:** It assumes the monitoring of microorganisms concentration before and after processing. The risk assessment systems are not using the microbiological criteria because the results are not obtained at an optimum time to initiate corrective/preventive actions.

2. **Thermal Resistance Characteristics:** The thermal resistance properties of the target microorganism are used as a model to forecast the influence of a specific heat treatment on the number of target microorganisms within a process. Two physical parameters (time and temperature) are monitored to calculate their impact on the number of microorganisms. The applicability of this method is limited as data that could accurately describe the temperature history of the studied microorganism is difficult to collect.

3. **Evaluation of Heat Treatments Effect:** Evaluation of the effect of heat treatments using mathematical models and time and temperature indicators. A time-temperature indicator (TI) is referred to as a small measuring component naturally present in a food matrix or added deliberately to food that can be quantified easily, precisely, and accurately to express the time, temperature history of a specific food (Mortier et al., 2000).

TI-s can be grouped based on principle action (biological, chemical, physical), response (simple, multiple), origin (intrinsic, extrinsic), and application (Claeys, 2003).

The milk components recommended for evaluation of heat treatments applied in the dairy industry as TI-s are related to enzymes like ALP, $\gamma$-GT, and LP, whey protein, such as $\beta$-lactoglobulin, chemical components as hydroxymethylfurfural, lactulose, and furosine. Enzymes meet the characteristics specified in the above definition. Temperature increase causes a rearrangement of the enzyme conformation with consequences on a catalytic activity whose loss can be analyzed easily, precisely, and accurately. This concept can be expressed by applying the standard method to describe the kinetics of thermal inactivation of a microorganism, according to the model decimal reduction time ($D$). The relationship between the $D$ indicator and the indicator enzyme target depends on the method sensitivity and concentration of the enzyme present in the food matrix.

To assess correctly the impact of thermal processes, the $z$ value (which is correlated to the resistance of a component under specific time/temperature conditions) of the indicator is recommended to be equal or higher than of the target indicator (Van Loey et al., 1995). This can be prevented by using experimental conditions in which the value of $z$ is known for both target and enzymatic indicator.

One common method to select a suitable indicator for the target indicator is to assess all the enzymes (in correlation with $D$ and $z$ value) contained by the investigated food matrix that will be heat treated (McKellar and Piyasena, 2000). The main disadvantage of using intrinsic indicators is the sensitivity to environmental factors, which limits their applicability. Another method consists in using the extrinsic indicators. This method requires the selection of an appropriate target indicator enzyme followed by an addition in the food system.

Hereinafter, the discussion will be focused on TI-s for pasteurization. The inactivation of heat-labile enzymes such as ALP, $\gamma$-GT, and LP will be evaluated to distinguish heat treatment of milk from different species.

### 1.2.1  ALKALINE PHOSPHATASE (ALP) (E.C. 3.1.3.1)

Milk contains a series of phosphatase, and from a technological point of view, alkaline and acid phosphomonoesterases are the most important. In milk, ALP is correlated with a milk fat globule membrane where more than 50% of its activity is associated with a fat fraction. ALP activity varies between species, breeds, and lactation cycle (Shakeel-Ur-Rehman, 2011). ALP possesses two main isozymes: $\alpha$- and $\beta$- phosphatase. $\alpha$- phosphatase is present in the milk plasma, while $\beta$-phosphatase is found intracellularly in the fat globule membrane. ALP is a hydrolase responsible for the dephosphorylation of nucleotides, proteins, and alkaloids. A number of features of ALP are shown in Table 1.2.

**TABLE 1.2**    ALP Characteristics in Cow Milk

| Characteristics | Conditions |
| --- | --- |
| Optimum pH | Casein 6,8; p-nitrophenyl phosphate 9,9 |
| Optimum temperature | 37°C |
| $K_m$ | 0.69 mM for p-nitrophenyl phosphate |

**TABLE 1.2** *(Continued)*

| Characteristics | Conditions |
|---|---|
| Activators | $Ca^{2+}$, $Mn^{2+}$, $Zn^{2+}$, $Co^{2+}$, $Mg^{2+}$ |
| Inhibitors | Chelators (EDTA), o-phosphate |
| Molecular weight | 170–190 kDa |
| Quaternary structure | Two subunits of 85 kDa |
| Zinc content | 4 mol.mol$^{-1}$ |
| Thermal stability | |
| D at 60°C, pH 9.0 | 27.2 min |
| D at 63°C, pH 9.0 | 8.3 min |

(*Source*: Reprinted with permission from Stănciuc, et al., 2011a. © Springer.)

### 1.2.1.1 IMPORTANCE

Milk is a rich source of nutrients for pathogenic bacteria, ALP test being used to establish thermal processing efficiency, or to identify the presence of raw milk in pasteurized milk products (Stănciuc et al., 2011c). ALP quantification is used to indicate suitable pasteurization conditions to milk. The non-spore-forming pathogenic bacteria such as Mycobacterium tuberculosis, Listeria monocytogenes are killed at temperatures lower than those necessary to inactivate ALP; therefore, the complete inactivation of ALP guarantees the safety consumption of milk.

McKellar et al. (1994) suggested that in specific conditions, ALP test is not accurate because: ALP partial reactivation may cause a false positive result; the enzyme is completed inactivated at time-temperature combinations less severe than classic HTST (72°C, 15 seconds); correlation between the logarithmic of its initial activity and pasteurization counterpart is less linear compared to other enzymes such as LP and γ-glutamyltranspeptidase.

### 1.2.1.2 ALP ACTIVITY DETERMINATION

ALP activity can be determined using immunochemical, colorimetric, fluorimetric, and chemiluminescent methods, out of which only the last three have been validated for milk pasteurization. In raw milk, the inorganic phosphate content increase during storage as a consequence of the hydrolysis of phosphoric esters by ALP. Commonly, at specific temperature conditions and pH, ALP catalyzes the hydrolysis of phosphoric monoesters, forming compounds that can be determined calorimetrically using phenyl phosphate or p-nitrophenyl phosphate as substrates. In general, the detection limit is equivalent to about 0.1% to 0.5% of the raw milk level (Clayes et al., 2003).

A more sensitive procedure is based on the use of commercial substrate *Fluorophos* (benzothiazoilhydroxy-benzothiazole phosphate). The detection limit of this method is equivalent to 0.003–0.006% in raw milk at excitation and an emission wavelength of 439 nm and 560 nm, respectively (ISO 11816–1:2013; IDF 155–1:2013). Another recent fluorimetric method for ALP assessment is based on the use of 8-quinolyl phosphate (QP). The fluorescent QP in the presence of ALP is converted to the nonfluorescent 8-hydroxyquinoline. Therefore, decreased fluorescence intensity via the enzyme reaction indicates a high ALP activity (Zhu and Jiang, 2007).

International Dairy Federation (IDF) and the International Organization for Standardization evaluated the reproducibility of chemiluminescent (Charm) method for ALP at 50, 100, 350, and 500 mU/mL in bovine, goat, sheep, and buffalo whole milk to meet new regulations in the US and proposed regulations in the EC. The results of this study revealed that the method was comparable to the fluorometric assays in performance, indicating that the chemiluminescent method is suitable for measuring ALP in the milk of multiple species. Charm method uses dioxen than phenyl phosphate as a substrate, which in the presence of ALP is converted to an unstable luminescent compound (ISO 22160, IDF 209:2007). According to Payne (2009), the detection limit is about 0.003% in raw milk.

The development of these sensitive techniques has set new limits for ALP activity in pasteurized milk that decreased from 500 mU/L to 350 mU/L; therefore, the classical colorimetric methods are currently considered unsuitable for ALP determination (Rankin et al., 2010).

### 1.2.1.3 LIMITATIONS

When assessing ALP activity, the influence of milk composition, reactivation capacity, and the presence of microbial ALP should be considered.

Painter and Bradley (1997) tested the ALP thermal stability in milk and cream and concluded that ALP residual activity is higher in milk with higher fat content. The presence of flavonoids and ascorbic acid inhibits ALP activity, NaCl 0.25 M decreases by about 50% the ALP thermal stability, while higher lactose content in milk increases the ALP thermal stability (Rankin et al., 2010). Moreover, the addition of additives in dairy products such as vanillin, sialic acid, or of antibiotics residues may lead to false-positive results. ALP reactivation is common where heat treatment involves higher temperatures and lower heating times (Fox and Kelly, 2006). The presence of $Mg^{2+}$, $Zn^{2+}$ ions, and NaCl stimulate ALP reactivation inducing conformational changes into denatured ALP structure, changes that are essential in the reactivation process (Figure 1.1). Also, sulfhydryl groups formed during heating at high temperatures may reactivate ALP, explaining the phenomenon for which ALP is reactivated only in UHT milk. Moreover, at high oxygen concentrations, the free sulfhydryl groups created during heating are later oxidized during storage, thus inhibiting sulfhydryl cross-linking reactions critical for the structure-dependent activity of ALP.

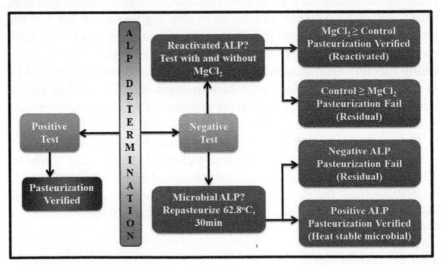

**FIGURE 1.1**   General logic to determine the pasteurization status of dairy products tested with the alkaline phosphate (ALP) assay.
(*Source*: Reprint from Rankin et al. (2010)).

## 1.2.2   γ-GLUTAMYLTRANSFERASE (GT) (EC 2.3.2.2)

The γ-GT is an enzyme present in body fluids and tissues of mammals, plants, and microorganisms. The enzyme is bound to the membranes, ~ 70% is found in skim milk, or bounded to the fat globule membrane. GT is involved in transport adjustment of cellular glutathione and amino acids through the glutamyl cycle, and in the biosynthesis of milk proteins. GT is a glycoprotein of ~80 kDa, having an optimum pH ranging between 8.5–9.0 at 45°C with an isoelectric point of 3.85. In milk, GT concentration varies between species with significant variations during the lactation period (Calamari et al., 2015). According to a study by Lorenzen et al. (2010), GT enzyme activity varies between 3420–5190 U/L in cow's milk, 1003–3133 U/L in sheep milk, and 335–1056 U/L in goat milk. About 25% of GT activity is distributed in the fat globules (Zehetner et al., 1995), and more than 70% of the total activity is found in the skim milk fraction (McKellar and Emmos, 1991). GT catalyzes the transfer of the γ-glutamyl group from peptides, such as glutathione, to other amino acids (Ziobro and McElroy, 2013). The enzyme operates on L-γ-glutamyl-p-nitroanilide, removing the glutamyl group to glycylglycine. The resulted p-nitroaniline is spectrophotometrically quantified at 410 nm (McKellar et al., 1991).

### 1.2.2.1   IMPORTANCE

GT is considered a useful indicator for assessing heat treatments applied to milk in the range between 70–80°C. The complete inactivation of GT occurs at 78°C/15 seconds or at 77°C/16 second. The inactivation is irreversible; the detection limit is about 0.1% and 0.25% raw milk in skim whole milk, respectively. Compared with ALP, GT is found in milk in higher concentrations possessing higher thermal stability than ALP. McKellar et al. (1996) suggested that due to the positive correlation between the reduction of GT activity and inactivation of streptococci, this enzyme is suitable to estimate heat treatments of milk between HTST and high pasteurization (>20 seconds at 85°C). Patel et al. (1994) suggested that fat content has no major influence on enzyme inactivation.

### 1.2.3   LACTOPEROXIDASE (LP) (EC 1.11.1.7)

LP belongs to the peroxidase (POD) family, being widely found in plants, but also present in animals and humans (Kussendrager and Van Hooijdonk, 2000). LP is a heme-containing glycoprotein that plays an important role in the natural host-defense systems. The antimicrobial and antiviral properties of the LP system in bovine milk are summarized by Seifu et al. (2005). LP is the second most abundant enzyme found in bovine milk (after xanthine oxidase), whose concentration ranges from 10 to 30 mg/L. Xanthine oxidase and superoxide dismutase are hydrogen peroxide sources, which improves the activity of the LP system.

### 1.2.3.1   REACTION MECHANISM OF LP SYSTEM

The reactions catalyzed by LP and response mechanisms are dependent on enzyme concentration, pH, and temperature. The LP system is able to convert thiocyanate to hypothiocyanite, in the presence of hydrogen peroxide (Şişecioğlu et al., 2010), which afterward react with microbial sulfhydryl groups to limit different cellular functions (Shin et al., 2002).

### 1.2.3.2   LACTOPEROXIDASE (LP) IN DAIRY INDUSTRY

LP is widely used in milk preservation. IDF has published the guidelines for milk handling and preservation using the LP-system (FAO, 1999). Antimicrobial agents from the LP system inhibit a wide range of pathogenic and spoilage bacteria, improving the microbiological quality of milk. Moreover, the activation of the LP system plays an important role in increasing the shelf life of raw and pasteurized milk (Trujilo et al., 2007). Sarkar and Misra (1994) indicated the factors that influence the utilization of LP-treated milk to obtain fermented milk products (Figure 1.2).

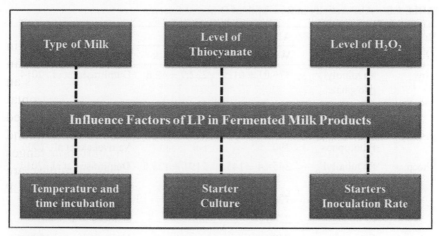

**FIGURE 1.2**    Influence factors of LP in fermented milk products.

## 1.3  KINETIC COMPARATIVE STUDIES ON INACTIVATION MECHANISMS OF ENDOGENOUS MILK ENZYMES FROM DIFFERENT SPECIES

### 1.3.1  KINETIC INVESTIGATION ON THERMAL INACTIVATION OF ALKALINE PHOSPHATASE (ALP)

#### 1.3.1.1  ALP ACTIVITY

In literature, the reported ALP activity for different types of milk varies widely (Table 1.3). Raynal-Ljutovac et al. (2007) suggested that differences are not correlated to species uniqueness but likely with milk fat fraction. Also, the difference in ALP activity may be related to a difference in ALP distribution. Raw cow milk contains higher contents of ALP than caprine milk and less than ovine milk. Martini et al. (2010) indicated that ALP of ovine milk is negatively associated to daily milk production and positively associated with milk fat fraction. Regardless of milk species, approximately 40% of ALP activity was found in the fat globule membrane. Therefore, one might expect that ALP from skim milk is less sensitive.

**TABLE 1.3**    ALP Activity in Raw Milk of Different Species

| Raw Milk Type | Method | ALP Activity | | References |
|---|---|---|---|---|
| | | Whole Milk | Skim Milk | |
| Cow | 8-Quinolyl phosphate | 376.01 ± 61.5 | 222.7 ± 49.8 | Dumitraşcu et al., 2014 |
| | Fluorophos | 328–1155 | *nm* | Lorenzen et al., 2010 |
| | *ns* | 330 | *nm* | Barbosa et al., 2005 |
| | Fluorophos | 390 | *nm* | Vamvakaki et al., 2006 |
| Sheep | 8-Quinolyl phosphate | 3458.4 ± 145.9 | 2191 ± 123.5 | Dumitraşcu et al., 2014 |
| | Fluorophos | 722–2691 | *nm* | Lorenzen et al., 2010 |
| | *ns* | 2200 | *nm* | Barbosa et al., 2005 |
| | Fluorophos | 2430 | *nm* | Vamvakaki et al., 2006 |
| Goat | 8-Quinolyl phosphate | 86.74 ± 2.9 | 52.4 ± 2.1 | Dumitraşcu et al., 2014 |
| | Fluorophos | 30–144 | *nm* | Lorenzen et al., 2010 |
| | Fluorophos | 134 | *nm* | Vamvakaki et al., 2006 |
| | *ns* | 95 | *nm* | Barbosa et al., 2005 |
| | *ns* | 128–236 | *nm* | Ying et al., 2002 |

*ns—not specified;*

*nm—not measured.*


## 1.3.1.2    THERMAL INACTIVATION MECHANISM OF ALP

The pasteurization efficacy by ALP assay in non-bovine milk was investigated by Lorenzen et al. (2010) and Wilinska et al. (2007). The authors measured in pasteurized milk samples ALP activity up to 62 mU/L, which is more than five times lower than the imposed legal limit for pasteurized milk. In the presence of some specific ions such as magnesium and zinc, the enzyme reactivates in UHT and pasteurized milk (Fox and Kelly, 2006; Lorenzen et al., 2011; Sosnowski et al., 2015).

On the other hand, in regard to milk from non-bovine species like goat and ewe, the limits have not been clearly defined. At the European level, the reference laboratory for milk and dairy products adopted the same limit for pasteurized goat milk as for cow milk of 350 mU/L and 10 mU/g for cheese made from pasteurized milk (Sosnowski et al., 2015). ALP thermal inactivation follows the first-order kinetic model (Claeys, 2002; Fadiloğlu et al., 2006; Stănciuc et al., 2011a; Dumitraşcu 2012) (Table 1.4).

**TABLE 1.4**    Comparative Analysis of Kinetic Parameters of ALP in Cow Milk

| Kinetic Parameters | Temperature (°C) | References | | | |
|---|---|---|---|---|---|
| | | Claeys, (2003) | Fadiloğlu et al. (2006) | Stănciuc et al. (2011a) | Dumitraşcu et al. (2012) |
| D (min) | 50°C | - | 217.0 ± 12.5 | 121.95 ± 20.23 | - |
| | 52.5°C | - | - | 55.86 ± 12.35 | - |
| | 54°C | 304.04 ± 24.85 | - | - | - |
| | 55°C | - | - | 49.5 ± 7.83 | - |
| | 56°C | 129.62 ± 5.65 | - | - | - |
| | 57.5°C | - | - | 42.19 ± 5.77 | - |
| | 58°C | 59.69 ± 2.89 | - | - | - |
| | 60°C | 24.62 ± 0.82 | 9.3 ± 0.27 | 27.02 ± 5.97 | 42.64 ± 3.97 |
| | 62°C | 9.65 ± 0.82 | - | - | |
| | 62.5°C | - | - | 23.09 ± 3.45 | 22.80 ± 1.98 |
| | 64°C | 4.32 ± 0.67 | - | - | |
| | 65°C | - | - | 14.77 ± 2.36 | 12.79 ± 0.52 |
| | 67.5°C | - | - | 13.83 ± 3.86 | 4.27 ± 0.15 |
| | 70°C | - | 2.9 ± 0.09 | 9.06 ± 2.59 | 1.88 ± 0.03 |
| | 72.5°C | - | - | 7.13 ± 1.35 | 0.85 ± 0.01 |
| | 75°C | - | - | 5.93 ± 0.93 | - |
| z (°C) | | 5.38 ± 0.06 | 11.8 ± 0.69 | 20.57 ± 0.84 | 7.15 ± 0.15 |
| Ea (kJmol/mol) | | 432.1 ± 6.3 | 207.8 ± 17.1 | 106.4 ± 4.76 | 311.70 ± 2.49 |

Wilinska et al. (2007) investigated at temperatures ranging from 54–69°C, the thermal inactivation of ALP in the whole goat, and cow milk using first order and Lumry-Eyring model. The authors concluded that a biphasic model described ALP inactivation with high accuracy at moderate temperatures and proposed a first-order kinetic model for the description of ALP inactivation at high temperatures. Biphasic model, as suggested by Dumitraşcu et al. (2014), indicated the presence of two fractions giving a complete overview of ALP thermal inactivation behavior in goat, cow, and sheep whole and skim milk. Also, Paintler and Bradly (1997) showed that ALP inactivation behavior follows two first-order reactions. The authors suggested the presence of two fractions with different thermal behavior inactivation. It seems that the fraction situated in milk fat is more heat resistant than that found in milk plasma. Dumitraşcu (2012) compared the inactivation kinetics of ALP in milk with and without fat, and no significant differences ($p>0.05$) were observed regardless of milk species.

Wilinska et al. (2007) reported $Ea$ values of 421 kJ/mol for cow milk and 406 kJ/mol for goat milk, whereas Stănciuc et al. (2011a) reported an $Ea$ value of $106.4 \pm 4.76$ kJ/mol for cow milk ALP (Table 1.4).

## 1.3.2 KINETIC INVESTIGATION ON THERMAL INACTIVATION OF GT

### 1.3.2.1 GT ACTIVITY

The highest GT activity was reported for sheep milk and the lowest for caprine milk (Figure 1.3). GT activity is correlated with fat content, as also reported for ALP. Enzymatic activity of GT was evaluated in correlation with protein, lactose, and fat content (Table 1.5). For cow milk, more than 20% of GT activity is correlated with fat content, whereas for GT of caprine origin, more than 50% of GT activity is related to fat globules. Piga et al. (2013) showed that the skimming procedure decreases the GT activity in sheep milk by about 35%. It seems that in the milk of non-bovine origin exists a stronger interaction between GT and lipid globules membranes. The same authors outlined the correlations between total concentrations of lipids, proteins, lactose, urea, and somatic cells and GT enzymatic activity. The GT activity reported by Lorenzen et al. (2010) for whole cow milk was 4143 U/L, 1878 U/L for sheep milk, and only 603 U/L for goat milk. During a lactation cycle, regardless of milk type, the authors haven't noticed significant variations between whole and skim milk. Piga et al. (2009) using chromatographic techniques reported for sheep milk (Sardinia breed) GT activity values between 2720 and 3460 U/L, while Stănciuc et al. (2011b) identified in whole and skim cow milk GT activity of 4010 and 3200 U/L respectively.

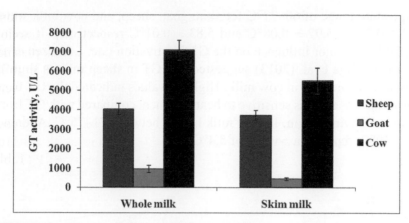

**FIGURE 1.3** GT activity in sheep, goat, and cow milk.

**TABLE 1.5** Medium Distribution of GT in Cow, Sheep, and Cow Milk

| Milk Type | Proteins (%) | Lactose (%) | Fat (%) | GT (U/L) | GT (%) | GT/Lactose Ratio |
|---|---|---|---|---|---|---|
| **Cow** | | | | | | |
| Whole | 3.43% | 4.50% | 3.96% | 7081 U/L | 100% | 1573 |
| Cream | 1.80% | 2.98% | 45% | 14845 U/L | 20.49% | 4981.54 |
| Skim | 3.53% | 4.61% | 0.30% | 5503.44 U/L | 77.71% | 1193.80 |
| **Sheep** | | | | | | |
| Whole | 4.68% | 6.17% | 8.39% | 4917 U/L | 100% | 796.92 |
| Cream | 2.20% | 3.1% | 50% | 11712 U/L | 23.9% | 3778 |
| Skim | 6.40% | 8.36% | 0.34% | 3727 U/L | 75.79% | 445.81 |
| **Goat** | | | | | | |
| Whole | 3.45% | 4.52% | 4.50% | 951 U/L | 100% | 210.39 |
| Cream | 1.7% | 2.67% | 47% | 5432 U/L | 51.12% | 2034.45 |
| Skim | 3.56% | 4.6% | 0.2% | 437 U/L | 45.95% | 95 |

(*Source*: Reprinted with permission from Dumitraşcu, 2012. © Elsevier.)

### 1.3.2.2  *THERMAL INACTIVATION MECHANISM OF GT*

GT inactivation follows a first-order kinetic (Figure 1.4). HTST pasteurization of cow milk resulted in the decrease of GT activity by about 54% and 28% for whole and skimmed milk, indicating that fat content influences the thermal inactivation behavior of GT (Hammershøj et al., 2010). Thermal sensitivity values ($z$) reported by Dumitraşcu et al. (2013) in the temperature range of 60–77°C for skim goat, sheep, and cow milk were: 8.02 ± 0.23°C, 5.97 ± 0.08°C and 5.83 ± 0.01°C respectively. It seems that fat has a minor influence on the GT inactivation rate. Lorenzen et al. (2010) and Piga et al. (2013) suggested that GT in sheep milk is slightly more heat-stable than in cow milk. Higher $z$-values indicated the GT from caprine milk as the less sensitive to heat treatment compared with GT from ovine and bovine origin. In cow milk heated between 71–75°C, Andrews et al. (1987) reported a $z$-value of 5.4°C.

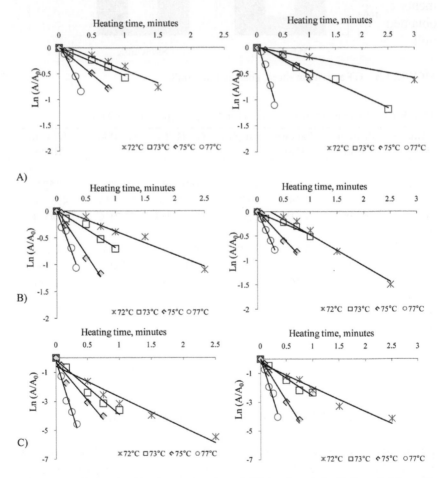

**FIGURE 1.4**   First-order kinetic inactivation of GT for the whole (left) and skim milk (right) at different temperatures. A) Sheep milk, B) Goat milk, and C) Cow milk. (*Source*: Modified from the data of Dumitrascu et al. (2012)).

## 1.3.3   KINETIC INVESTIGATION ON THERMAL INACTIVATION OF LP

### 1.3.3.1   LP ACTIVITY

An assay of LP activity is used by Griffiths (1986) as a useful procedure to identify heat pasteurization treatments performed at 76°C for 15 seconds, while Fox (2003) concluded that LP may be used to evaluate heat treatments above 72°C, 15 seconds, total inactivation of this enzyme being obtained at 78°C for 15 seconds. LP activity was measured mainly in cow milk, limited studies being available for LP of non-bovine origin. In literature, various data were reported for LP activity; these differences being associated with animal welfare, breed, season, species, etc. (Table 1.6).

**TABLE 1.6**   Lactoperoxidase Activity in Raw Milk of Different Species

| Type of Raw Milk | Breed | LP Activity (U/mL) | References |
|---|---|---|---|
| Cow | Na* | 1.2–19.4 U/mL | Seifu et al., 2005 |
| | Romanian Simmental | 0.97 ± 0.012 U/mL | Dumitraşcu et al., 2012 |
| | Friesian | 6.1–6.9 U/mL | Turner et al., 2007 |
| Goat | Alpine Saanen | 1.73 U/mL | Trujillo et al., 2007 |
| | White Banat | 0.81 ± 0.0 U/mL | Dumitraşcu et al., 2012 |
| | Saanen | 0.21 U/mL | Parry-Hanson et al., 2009 |
| | Saanen | 0.79 ± 0.18 U/mL | Seifu et al., 2004b |
| Sheep | Na | 0.14–2.38 U/mL | Medina et al., 1989 |
| | Machenga | 1.38–6.10 U/mL | Althaus et al., 2001 |
| | Merino | 1.72 ± 0.05 U/mL | Dumitraşcu et al., 2012 |

*Information not available.

Dumitraşcu et al. (2012) reported an average LP activity of 1.72 U/mL in sheep milk, 0.81 U/mL in goat milk, and 0.97 U/mL in cow milk. Chavarri et al. (1998) measured in sheep milk LP activities between 8.65 ± 0.38–10.57 ± 0.34 U/mL.

### 1.3.3.2   THERMAL INACTIVATION MECHANISM OF LP

Dumitraşcu et al. (2012) examined the thermal stability of LP (using ABTS method) in milk from bovine and non-bovine species at temperatures up to 77°C. For whole tested temperatures, the highest residual enzyme activity was recorded in cow milk and the lowest in the milk of caprine origin. After 5 minutes of heating at 73°C, the residual activity of LP decreased by about 56% in goat milk, 40% in sheep milk, and 35% in cow milk (Figure 1.5).

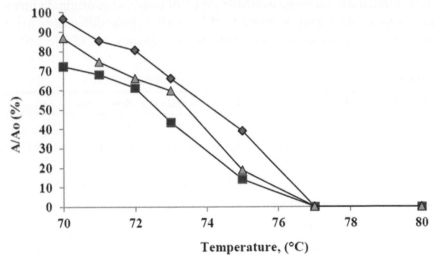

**FIGURE 1.5**   Thermal stability of LP in cow milk (blue), goat milk (red), and in sheep milk (green).
(*Source*: Reprint from Dumitrascu et al. (2012)).

Thermal inactivation of LP was accurately described by first-order; the kinetics parameters indicated that LP is more sensitive in non-bovine milk at temperatures below 75°C. Lorenzen et al. (2010) indicated that heating at 75°C, 28 seconds decreased the LP activity to about 40%, the decrease being independent of milk origin. The kinetic parameters for thermal inactivation of LP in different types of milk show significant differences between species (Table 1.7). Clayes (2003) concluded that LP thermal inactivation is not influenced by milk fat content. Also, the same author indicated that complete inactivation of LP takes place at conditions less severe than those associated with high pasteurization, suggesting the use of this enzyme for defining the intermediate pasteurization.

**TABLE 1.7**  Kinetic Parameters for First Order Thermal Inactivation of LP in Different Milk Species

| Milk Type | Temperature Range (°C) | z(°C) | References |
|---|---|---|---|
| Cow | 70–80°C | 3.58 ± 0.004°C | Dumitraşcu et al., 2012 |
| | 69–73°C | 3.74 ± 0.31°C | Claeys, 2002 |
| | 67–73°C | 4.7°C | Tayefi-Nasrabadi et al., 2011 |
| Goat | 73–78°C | 3.7°C | Olszewski and Reuter, 1992 |
| | 70–80°C | 3.38 + 0.01°C | Dumitraşcu et al., 2012 |
| Sheep | 69–73°C | 9.45°C | Trujilo et al., 2007 |
| | 70–80°C | 4.11 ± 0.24°C | Dumitraşcu et al., 2012 |

## 1.4  CONCLUSIONS

The purpose of this chapter was to evaluate to what extent variation in milk composition affects the inactivation thermal behavior of alkaline phosphates, $\gamma$-GT, and LP. Enzymatic activity and inactivation kinetics of the selected compounds was evaluated in cow, ovine, and caprine milk. Enzymatic activities in raw milk are dependent on milk species. For all the studied enzymes, the highest enzymatic activities were identified in bovine milk, and the lowest in caprine milk. In regard to ALP, about 40% of its activity is correlated with fat content, without variations between species. Association of milk fat content with $\gamma$-GT activity indicated important variations between bovine and nonbovine milk species. Inactivation kinetics of the tested enzymes fitted the first-order kinetic model, while the fat content did not affect the enzyme inactivation behavior. Based on data reported, alkaline phosphates, $\gamma$-GT seems to be suitable markers for milk pasteurization, while LP might be considered a good indicator for marking the region of intermediate pasteurization for both bovine and nonbovine milk tested. Therefore, the application of several enzymatic indicators is useful to define the time-temperature combinations to distinguish heat treatment applied to milk.

## KEYWORDS

- alkaline phosphatase
- International Dairy Federation
- lactoperoxidase
- non-bovine milk
- peroxidase
- γ-glutamyltransferase

## REFERENCES

Al-Hilphy, A. R. S., Verma, D. K., Niamah, A. K., Billoria, S., & Srivastav, P. P., (2016). Principles of ultrasonic technology for treatment of milk and milk products.In: Meghwal, M., Goyal, M. R., (eds.), *Food Process Engineering: Emerging Trends in Research and Their Applications* (Vol. 5). As part of book series on "Innovations in Agricultural and Biological Engineering", Apple Academic Press, USA.

Althaus, R. L., Molina, M. P., & Rodriguez, M., (2001). Analysis time and lactation stage influence on lactoperoxidase system components in dairy ewe milk. *Journal of Dairy Science, 84*, 1829–1835.

Andrews, A. T., Anderson, M., & Goodenough, P. W., (1987).A study of heat stabilities of a number of indigenous milk enzymes. *Journal of Dairy Research, 54*, 237–246.

Barbosa, M., (2005).Interest in controlling alkaline phosphatase activity in sheep and goat milks. In: *Proc. International Symposium, Zaragoza* (Special Issue, pp. 117–121). Spain. International Dairy Federation, Brussels, Belgium.

Calamari, L., Gobbi, L., Russo, F., & Cappelli, F. P., (2015). Pattern of gamma-glutamyltransferase activity in cow milk throughout lactation and relationships with metabolic conditions and milk composition.*Journal of Animal Science, 93*(8), 3891–3900.

Chavarri, F., Santistebam, A., Virto, M., & De Renobales, M., (1998). Alkaline phosphatase, acid phosphatase, lactoperoxidase and lipoprotein lipase activities in industrial ewe's milk and cheese. *Journal of Agricultural and Food Chemistry, 46*, 2926–2932.

Claeys, W. L., Cardoen, S., Daube, G., De Block, J., Dewettinck, K., Dierick, K., & Herman, L., (2013). Raw or heated cow milk consumption: Review of risks and benefits. *Food Control, 31*(1), 251–262.

Claeys, W. L., Verraes, C., Cardoen, S., De Block, J., Huyghebaert, A., Raes, K., & Herman, L., (2014). Consumption of raw or heated milk from different species: An evaluation of the nutritional and potential health benefits. *Food Control, 42*, 188–201.

Claeys, W., (2003). Intrinsic time temperature integrators for thermal and high pressure processing of milk. *PhD Thesis*. KatholiekeUniversiteit Leuven. Belgium.

Claeys, W., Van Loey, A., & Hendricks, M., (2002). Kinetics of alkaline phosphatase and lactoperoxidase inactivation, and of β-lactoglobulin denaturation in milk with different fat content.*Journal of Dairy Research*, *69*, 541–553.
Add:
Dumitrascu, L.; Stănciuc, N.; Stanciu, S.; Râpeanu, G. Thermal inactivation of lactoperoxidase in goat, sheep and bovine milk – A comparative kinetic and thermodynamic study, Journal of Food Engineering, 2012, 113 (1), 47–52.
Fadiloğlu, S., Erkmen, O., & Şekeroğlu, G., (2006). Thermal inactivation kinetics of alkaline phosphatase in buffer and milk. *Journal of Food Processing and Preservation*, *30*(3), 258–268.
FAO, (1999). *Manual on the Use of the LP System in Milk Handling and Preservation*. Rome: Food and Agriculture Organization of the United Nations.
Fox, P. F., & Kelly, A. L., (2006). Indigenous enzymes in milk: Overview and historical aspects—Part 2, *International Dairy Journal*, *16*, 517–532.
Fox, P. F., (2003). Significance of indigenous enzymes in milk and dairy products. In: Whitaker, J. R., Voragen, A. G. J., & Wong, D. W. S., (eds.), *Handbook of Food Enzymology*. Marcel Dekker, Inc. New York.
Gallusser, A., & Bergner, K. G., (1981). Reactivation of alkaline phosphatase in UHT milk, depending on its oxygen content.*Deut. Lebensm. Rundsch.*,*77*, 441–444.
Griffiths, M. W., (1986). Use of milk enzymes as indices of heat treatment. *Journal of Food Protection, 49*, 696–705.
Hammershøj, M., Hougaard, A., Vestergaard, J., Poulsen, O., & Ipsen, R., (2010). Instant infusion pasteurization of bovine milk. ii. Effects on indigenous milk enzymes activity and whey protein denaturation. *International Journal of Dairy Technology, 63*, 2, 197–208.
ISO 11816–1:2013 (IDF 155–1:2013). *Milk and Milk Products – Determination of Alkaline Phosphatase Activity -- Part 1: Fluorimetric Method for Milk and Milk-Based Drinks*.
Kussendrager, K. D., & Van Hooijdonk, A. C. M., (2000). Lactoperoxidase: Physicochemical properties, occurrence, mechanism of action and applications. *British Journal of Nutrition, 84*(1), S19–S25.
Lorenzen, P. Chr., Martin, D., Clawin-Radecker, I., Barth, K., & Knappstein, K., (2010). Activities of alkaline phosphatase, γ-glutamyltransferase and lactoperoxidase in cow, sheep and goat's milk in relation to heat treatment.*Small Ruminant Research*, 89, 1, 18–23.
Marchand, S., Merchiers, M., Messens, W., Coudijzer, K., & De Block, J., (2009).Thermal inactivation kinetics of alkaline phosphatase in equine milk.*International Dairy Journal*, *19*, 763–767.
Martini, M., Salari, F., Pessi, R., & Tozzi, M. G., (2010). Relationship between activity of some fat globule membrane enzymes and the lipidic fraction in ewes' milk: Preliminary studies. *International Dairy Journal, 20*(1), 61–64.
McKellar, R. C., (1996). Influence of ice-cream mix components on the thermal stability of bovine milk γ-glutamyltranspeptidase and *Listeria innocua*.*International Dairy Journal,6*(11&12), 1181–1189.
McKellar, R. C., Emmons, D. B., & Farber, J., (1991).Gamma glutamyltranspeptidase in milk and butter as an indicator of the heat treatment. *International Dairy Journal, 1*, 241–251.

McKellar, R. C., Modler, H. W., Couture, H., Hughes, A., Mayers, P., Gleeson, T., & Ross, W. H., (1994). Predictive modeling of alkaline phosphatase inactivation in a high-temperature short-time pasteurizer. *Journal of Food Protection, 57*, 424–430.

McKellar, R., & Piyasena, P., (2000). Predictive modeling of inactivation of bovine milk α-L fucosidase in high temperature short time pasteurizer. *International Dairy Journal, 10*, 1–6.

Medina, M., Gaya, P., & Nunez, M., (1989). The lactoperoxidase system in ewe's milk: Levels of lactoperoxidase and thiocyanate. *Letters in Applied Microbiology, 8*, 147–149.

Moatsou, G., (2010). Indigenous enzymatic activities in ovine and caprine milks. *International Journal of Dairy Technology, 63*(1), 16–31.

Mortier, L., Braekman, A., Cartuyvels, D., Renterghem, R., & De Block, J., (2000). Intrinsic indicators for monitoring heat damage of consumption milk. *Biotechnological Agronomical Society Environment, 4*, 221–225.

Murthy, G. K., Kleyn, D. H., Richardson, T., & Rocco, R. M., (1992). Alkaline phosphatase methods. In: Marshall, R. T., (ed.), *Standard Methods for the Examination of Dairy Products* (pp. 413–431). Am. Publ. Health. Assoc., Washington, DC.

Niamah, A. K., & Verma, D. K., (2017). Microbial intoxication in dairy food product. In: Verma, D. K., & Srivastav, P. P., (eds.), *Microorganisms in Sustainable Agriculture, Food and the Environment* (Vol. 1). as part of book series on "Innovations in Agricultural Microbiology", Apple Academic Press, USA.

Olszewski, E., & Reuter, H., (1992). The inactivation and reactivation behavior of lactoperoxidase in milk at temperature between 50°C and 135°C. *Zeitschriftfür Lebensmittel-Untersuchung und Forschung, 194*, 235–239.

Painter, C. J., R. L. Bradley, Jr., (1997). Residual alkaline phosphatase activity in milks subjected to various time-temperature treatments. *Journal of Food Protection, 60*, 525–530.

Parry-Hanson, A., Jooste, P. J., & Buys, E. M., (2009).The influence of lactoperoxidase, heat and low pH on survival of acid-adapted and non-adapted *Escherichia coli O157:H7* in goat milk. *International Dairy Journal, 19* 417–421.

Patel, S., & Wilbey, R., (1994). Thermal inactivation of γ-glutamyltranspeptidase and *Enterecoccusfaecium* in milk-based systems. *Journal of Dairy Research, 61*, 263–270.

Piga, C., Urgeghe, P. P., Piredda, G., Scintu, M. F., & Sanna, G., (2009). Assessment and validation of methods for the determination of γ-glutamyltransferase activity in sheep milk. *Food Chemistry, 4*, 1519–1523.

Piga, C., Urgeghe, P., Piredda, G., Scintu, M. F., Di Salvo, R., & Sanna, G., (2013). Thermal inactivation and variability of g-glutamyltransferase and α L-fucosidase enzymatic activity in sheep milk. *LWT – Food Science and Technology,54*, 152–156.

Rankin, S. A., Christiansen, A., Lee, W., Banavara, D. S., & Lopez-Hernandez, A., (2010). Invited review: The application of alkaline phosphatase assays for the validation of milk product pasteurization. *Journal of Dairy Science, 93*, 5538–5551.

Raynal-Ljutovac, K., Park, Y. W., Gaucheron, F., & Bouhallab, S., (2007). Heat stability and enzymatic modifications of goat and sheep milk. *Small Ruminant Research, 68*(1&2), 207–220.

Regulation (EC) 853/2004 of the European Parliament and of the Council of 29 April 2004 laying down specific hygiene rules for on the hygiene of foodstuffs. *Official Journal of the European Union*.

Salter, R. S., & Fitchen, J., (2006). Evaluation of a chemiluminescence method for measuring alkaline phosphatase activity in whole milk of multiple species and bovine dairy drinks: Inter laboratory study. *J. AOAC Int.*, *89*, 1061–1070.

Sarkar, S., & Misra, A. K., (1994a). Milk preservation by LP system and its effect on the quality of pasteurized milk. *Indian Journal of Dairy Science*, *47*, 780–784.

Seifu, E., Buys, E. M., & Donkin, E. F., (2005). Significance of lactoperoxidase system in the dairy industry and its potential applications: A review. *Trends in Food Science and Technology*, *16*, 137–154.

Seifu, E., Buys, E. M., Donkin, E. F., & Petzer, I. M., (2004b). Antibacterial activity of the lactoperoxidase system against foodborne pathogens in Saanen and South African indigenous goat milk. *Food Control*, *15*, 447–452.

Shakeel-Ur-Rehman, C. M., Fleming, N. Y., & Farkye, F. P. F., (2003).Indigenous phosphatases in milk. In: Fox, P. F., & McSweeney, P. L. H., (eds.), *Advanced Dairy Chemistry: Proteins* (3rd edn., Vol. 1, pp. 523–544). Kluwer Academic/Plenum Publishers, New York, NY.

Shin, K., Yamauchi, K., Teraguchi, S., Hayasawa, H., & Imoto, I., (2002). Susceptibility of helicobacter pylori and its urease activity to the peroxidase-hydrogen peroxide-thiocyanate antimicrobial system. *Journal of Medical Microbiology*, *51*, 231–237.

Şişecioğlu, M., Çankaya, M., Gülçin, İ., & Özdemir, H., (2010). Interactions of melatonin and serotonin to lactoperoxidase enzyme. *Journal of Enzyme Inhibition and Medicinal Chemistry*, *25*, 779–783.

Slacanac, V., Bozanic, A., Judit, H. J., Szabo, E., Lucan, M., & Krstanovic, V., (2010). Nutritional and therapeutic value of fermented caprine milk. *International Journal of Dairy Technology*, *63*(2), 171–189.

Sosnowski, M., Rola, J. G., & Osek, J., (2015). Alkaline phosphatase activity and microbiological quality of heat-treated goat milk and cheeses. *Small Ruminant Research*. http://dx.doi.org/10.1016/j.smallrumres.2015.12.038 (Accessed on 8 November 2019).

Stănciuc, N., Ardelean, A., Diaconu, V., Râpeanu, G., Stanciu, S., & Nicolau, A., (2011a). Kinetic and thermodynamic parameters of alkaline phosphatase and γ – glutamyltransferase inactivation in bovine milk. *Dairy Science and Technology*, *91*(6), 701–717.

Stănciuc, N., Dumitrascu, L., Râpeanu, G., & Stanciu, S., (2011b). γ-Glutamyltransferase inactivation in milk and cream: A comparative kinetic study. *Innovative Food Science and Emerging Technologies*, *12*, 56–61.

Stănciuc, N., Râpeanu, G., & Stanciu, S., (2011c). Traceability. In: Academica, (ed.), *Fundamental Concepts Specific to Milk and Dairy Industry (Written in Romanian)* (p. 270). ISBN: 978–973–8937–73–4.

Tayefi-Nasrabadi, H., Hoseinpour-Fayzi, M. A., & Mohasseli, M., (2011). Effect of heat treatment on lactoperoxidase activity in camel milk: A comparison with bovine lactoperoxidase. *Small Ruminant Research*, *99*(2), 187–190.

Trujillo, A. J., Pozo, P. I., & Guamis, B., (2007). Effect of heat treatment on lactoperoxidase activity in goat milk. *Small Ruminant Research*, *67*, 243–246.

Turner, S. A., Thomson, N. A., & Auldist, M. J., (2007). Variation of lactoferrin and lactoperoxidase in bovine milk and the impact of level of pasture intake, New Zealand. *Journal of Agricultural Research*, *50*(1), 33–40.

Vamvakaki, A., Zoidou, E., Moatsou, G., Bokan, M., & Anifantakis, E., (2006).Residual alkaline phosphatase activity after heat treatment of ovine and caprine milk. *Small Ruminant Research*, *65*(3), 237–241.

Van Loey, A., Hendrickx, M., De Cort, S., Haentjens, T., & Tobback, P., (1996). Quantitative evaluation of thermal processes using time temperature integrators. *Trend in Food Science and Technology*, *7*, 16–25.

Van Loey, A., Ludikhuyze, L., Hendrickx, M., De Cort, S., Haentjens, T., & Tobback, P., (1995). Theoretical consideration on the influence of the z-value of a single component time/temperature integrator on thermal process impact evaluation. *Journal of Food Protection*, *58*, 39–48.

Verma, D. K., Mahato, D. K., Billoria, S., Kapri, M., Prabhakar, P. K., Ajesh, K. V., &Srivastav, P. P., (2017). Microbial spoilage in milk products, potential solution, food safety and health issues. In: Verma, D. K., & Srivastav, P. P., (eds.), *Microorganisms in Sustainable Agriculture, Food and the Environment* (Vol. 1). As part of book series on "Innovations in Agricultural Microbiology", Apple Academic Press, USA.

Wilińska, A., Bryjak, J., Illeová, V., & Polakovič, M., (2007). Kinetics of thermal inactivation of alkaline phosphatase in bovine and caprine milk and buffer. *International Dairy Journal*, *17*, 579–586.

Ying, C., Yang, C. B., & Hsu, J. T., (2002). Relationship of somatic cell count, physical, chemical and enzymatic properties to the bacterial standard plate count in different breads of dairy goats. *Livestock Production Science*, *74*, 63–77.

Zehetner, G., Bareuther, C., Henle, T., & Klostermeyer, H., (1995). Inactivation kinetics of $\gamma$-glutamyl-transferase during the heating of milk. *Zeitschriftfür Lebensmittel-Untersuchung und-Forschung, 201*, 336–339.

Zhu, X., & Jiang, C., (2007). 8-Quinolyl phosphate as a substrate for the fluorimetric determination of alkaline phosphatase. *Clinica Chimica Acta*, *377*, 150–153.

Ziobro, G., & McElroy, K., (2013). Fluorometric detection of active alkaline phosphatase and $\gamma$-glutamyltransferase in fluid dairy products from multiple species. *Journal of Food Protection*, *76*(5), 892–898.

# CHAPTER 2

# Microwave Heating: Alternative Thermal Process Technology for Food Application

DEEPAK KUMAR VERMA,[1] NAVEEN KUMAR MAHANTI,[2]
MAMTA THAKUR,[3] SUBIR KUMAR CHAKRABORTY,[4] and
PREM PRAKASH SRIVASTAV[1]

[1]*Agricultural and Food Engineering Department,
Indian Institute of Technology, Kharagpur, West Bengal, India,
E-mails: deepak.verma@agfe.iitkgp.ernet.in; rajadkv@rediffmail.com
(D.K. Verma); E-mail: pps@agfe.iitkgp.ernet.in (P.P. Srivastav)*

[2]*Agricultural Processing and Structures Division, Indian Agricultural
Research Institute, New Delhi, India, E-mail: naveeniitkgp13@gmail.com*

[3]*Department of Food Engineering and Technology, Sant Longowal
Institute of Engineering and Technology, Longowal, Punjab, India,
E-mails: thakurmamtafoodtech@gmail.com; mamta.ft@gmail.com*

[4]*Agro Produce Processing Division, ICAR-Central Institute of
Agricultural Engineering, Bhopal, Madhya Pradesh, India,
E-mail: Subir.Kumar@icar.gov.in*

## 2.1 INTRODUCTION

Microwaves refer to the electromagnetic waves having frequency varying from 300 MHz to 300 GHz and wavelength from 1 mm to 1 m (Okeke et al., 2014; Chandrasekaran et al., 2013; Benlloch-Tinoco et al., 2013; Guan et al., 2004; Venkatesh and Raghavan, 2004). The food industry utilizes the only limited frequencies because of the interaction between the microwaves and telecommunication devices. Generally, the industrial and domestic microwave ovens employ the 915 and 2450 MHz frequencies, respectively (Sumnu and Sahin, 2015). However, the microwave power usage has been quickly raised in the food industry due to its positive impacts over the conventional method in terms of minimum processing time and energy requirement, precise control and instant startup and shutdown times (Kidmose and Martens, 1999). The heat energy is formed during the microwave application from electromagnetic energy through the process of selective absorption and dissipation (Guo et al., 2017). Further, the potential of microwave conversion into heat is also affected by the dielectric properties of food that are the major factors which determine the electromagnetic energy coupling and distribution. This is expressed on the basis of dielectric constant/relative permittivity and dielectric loss (Wang et al., 2003; Guo et al., 2017) where the dielectric constant relates the potential of a substance to polarize and store the electrical energy in response to applied electric field while the dielectric loss deals with the energy dissipation in the form of heat (Datta et al., 2005; Sosa-Morales et al., 2010; Franco et al., 2015). Usually, the dielectric characteristics rely on the composition, salt, fat, moisture content *(MC)*, and temperature of food and applied frequency (Tang, 2015). But, microwave heating results in the non-uniform distribution of heat because the microwave hotspot pattern is generated at such a point where the electromagnetic field intensity is higher, and there is a rapid transfer of energy while employing the high frequencies of microwaves. This quick heating maintains the product quality after drying also, but results in the uneven heating of the end product (Guo et al., 2017).

The microwaves, when interacts with the food constituents, generate heat due to the molecular friction and excitation (Sun et al., 2007). As compared to conventional heating, the microwave heating is 3–5 times faster, and also the mode of heat and mass transfer is different, which improves the product quality (Sun et al., 2007). The heat transfers from outside of product to inside in conventional heating system; however, there is heat movement from inside to outside of product in microwave heating. The mechanism of heat and mass transfer of microwave heating is described in detail in the further sections of the chapter. The potential of microwaves in food processing industries is shown in Figure 2.1.

Thus the chapter entitled "Microwave Heating: Alternative Thermal Process Technology for Food Application" focuses on the principle of a microwave oven, the heat, and mass transfer mechanism, comparison of conventional and microwave heating, factors affecting the microwave heating, advantages of microwaves in food processing operations.

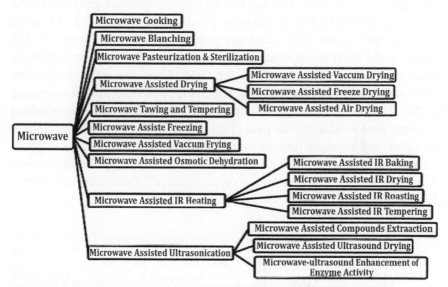

**FIGURE 2.1** Potential of microwave in the food industry (Chizoba et al., 2017).

## 2.2 MICROWAVES: AN OVERVIEW

Microwaves, as discussed above, have a frequency from 300 MHz to 300 GHz, where the product heating occurs on the basis of microwaves absorption by dielectric materials. During the process, the heat is generated at the product center and shifted towards the surface, which is not observed in conventional heating. The ease for cleaning, volumetric heating (process of interaction of electromagnetic field and with food) and quick processing thus maintain the quality are associated with microwave heating (Sumnu and Sahin, 2005; Ahmed and Ramaswamy, 2007) whereas the higher processing time (due to slow thermal diffusion process) reduces the product quality in traditional pasteurization and retorting techniques. Generally, the microwave heating requires 1/4th time to reach processing temperature as compared to conventional process (Tewari, 2007).

### 2.2.1 INSTRUMENTATION AND MECHANISM OF WORKING PRINCIPLE

During microwave processing, the dielectric material generates heat due to the rotation of dipole and ionic polarization (Figure 2.2). Usually, the polar molecules like water are present in all food materials. On exposing the food to an electric field, there is an orientation of polar molecules as per the polarity of the field. Then the water molecules start reorientation due to the repeated changes in field polarity, leading to the friction generation between the molecules and surrounding medium, thus producing the heat (Figure 2.2A). On the other side, there are collisions between ions in ionic polarization technique resulting in the heat on exposing the food to an electric field due to the movement of ions at an accelerated phase on account of their inherent charge (Figure 2.2B).

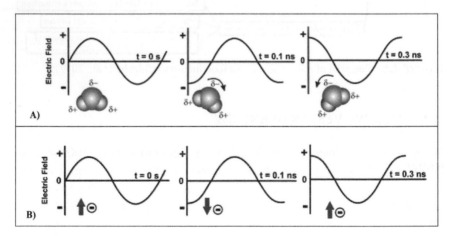

**FIGURE 2.2**   The detailed mechanism of heat generation in microwave heating.(A) Dipole rotation, and (B) Ionic conduction.
(*Source*: Reprinted with permission from Bilecka and Niederberger, 2010. © Royal Society of Chemistry.)

The main parts of the microwave oven are MW source, a waveguide, and applicator. The heart of the microwave oven is a magnetron (Figure 2.3); it consists of a vacuum tube with a central electron-emitting cathode surrounded by an anode (Regier et al., 2010). When high voltage current passes through magnetron, it converts into MW energy and emits high-frequency radiant energy. The polarity of radiant energy changes from positive to negative at high frequencies. The magnetron power usually varies from 300–3000 W, but different magnetrons are used in industrial microwaves in order to increase the power output. The heating chamber holds the product for heating due to the microwaves as guided by the microwave guide. Commercially, the microwave frequencies 2450 and 915 MHz are used for heating food, where 2450 MHz is used for home ovens (Regier et al., 2010).

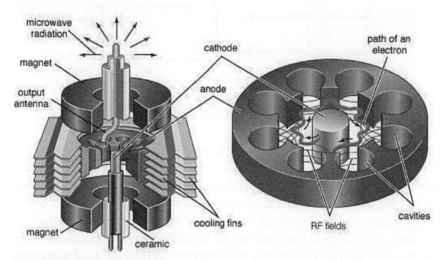

**FIGURE 2.3**   Magnetron and its parts (*Source*: Reprinted from Narang and Gupta, 2015).

## 2.3   MECHANISM OF HEAT AND MASS TRANSFER

The conventional and microwave heating differentiates from each other on the basis of heat transfer direction. The heat flux first reaches the product surface and then transferred to the center of the product due to the thermal gradient in conventional drying. Here, the product surface on prolonged exposure to higher temperature results in the shrinkage cracking, and sometimes, case hardening. The generation of a hard heat-resistant layer above the product surface during heating is referred to as case hardening, which ultimately leads to a reduction in the drying rate, thus deteriorating the product quality and increasing the drying time also. Such a solid dry part also restrains the transfer of water from the product center. Hence, there is a common issue of still wet solid areas within the theoretically dry material. The increase in temperature inside the food is fixed while the drying process, which is independent of *MC*. During conventional heating, the direction of the flow of heat mass is exhibited in Figure 2.4.

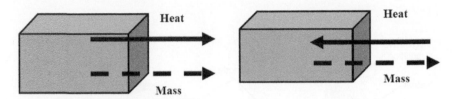

**FIGURE 2.4** The mechanism of heat and mass transfer in A) microwave and B) conventional heating (Araszkiewicz et al., 2007).

On the other hand, the microwave energy is transferred directly to the food material through the process of molecular interaction with the electromagnetic field resulting in the generation of heat within the food molecules, and the highest temperature is observed at the food center which dries faster compared to the surface. The distinct feature of microwave heating is the volumetric heating in which the food material absorbs microwave energy and converts it into the heat generated throughout the product (Non-uniform temperature). In this case, the heat generation rate depends on the water content of food (Puffing-Microwave). The method of delivering energy using microwaves may result in numerous potential advantages for food processing (Venkatesh and Raghavan, 2004).

## 2.4   FACTORS AFFECTING MICROWAVE HEATING

There are various factors that affect microwave heating and they are listed in Table 2.1.

**TABLE 2.1**   List of Factors Influencing the Dielectric Characteristics

| Factor(s) | Effect(s) |
| --- | --- |
| Frequency | The polarization results in the dependence of dielectric properties on frequency due to the dipoles orientation using the applied field. As the frequency increases, the dielectric constant reduces or remains constant while other conditions are constant. The dielectric characteristics are based on the frequency range and absorption. The frequency, when increased beyond the limit generates no dipole motion but changes the field leading to the reduction of dielectric constant and the absorption of energy due to the phase lag between dipole rotation and field. Here, the ionic conductivity is highly significant at lower frequencies (<200 MHz), whereas the ionic conductivity and dipole rotation of free water has a synergistic role at higher microwave frequencies (like 915 and 1800 MHz). |
| Moisture | The kind of water found in food decided the dielectric properties. Generally, the free water has pronounced effect compared to bonded water. The higher the moisture, there is an increase in dielectric constant and loss factor. However, in some cases, the loss factor reduces when moisture is increased at some frequency ranges. Feng et al. reported the increase in the dielectric constant of apple with an increase in temperature over the frequency range (915–1800 MHz) when moisture is below 70% but decreases when moisture is greater than 70%. |
| Temperature | The higher temperature results in water evaporation. On reduction of moisture content, the dielectric loss, loss tangent, and dielectric constant are affected. Increased temperature raises the water mobility resulting in dielectric properties. The increase or decrease of the dielectric loss factor is based on the applied frequency compared to the relaxation frequency. Dielectric loss factor due to the ionic conductivity increases, whereas the loss factor due to dipole rotation of free water reduces with the temperature rise at particular frequencies (like 915 MHz and 2450 MHz). |
| Density | The highly porous food material has low dielectric properties due to the extremely low dielectric properties of air. The dielectric constant of air is 1 while the loss factor is 0. The dielectric properties are further reduced while baking of bread in the microwave oven further reduction of moisture due to the formation of pores. There is a higher effect of porosity on the dielectric properties compared to moisture content. |
| Salt | The permittivity of foods is reduced due to the presence of dissolved salts. The loss factor usually increases due to the presence of salts as a conductor in electromagnetic fields. |

(*Sources*: Modified from: Heddleson and Doores, (1994) ; Zhang and Datta, (2001); Feng et al. (2002); Wang et al. (2003); Icier and Baysal, (2004); Rahman, (2007); Franco et al. (2015)).

### 2.4.1   FREQUENCY

The penetration depth is directly related to the frequency of microwaves; frequently used frequencies are 915 and 2450 MHz, corresponding to the wavelengths 0.328 m and 0.122 m, respectively (Ahmed and Ramaswamy, 2007). The depth at which power is reduced to 1/e is referred to power penetration depth which is calculated using the following equation:

$$d_p = \frac{\lambda_0}{2\pi\sqrt{2\varepsilon'}} \left[ \sqrt{1 + \left(\frac{\varepsilon''}{\varepsilon'}\right)^2} - 1 \right]^{\frac{-1}{2}}$$

where:

$d_p$ = Depth of penetration (m);
$\lambda_0$ = Electromagnetic wavelength in free space (m) = $c_0/f$;
$c_0$ = Light speed in free space ($2.9979 \times 10^8$ m/s);
$f$ = Field frequency (Hz);
$\varepsilon'$ = Dielectric constant (or) relative electrical permittivity;
$\varepsilon''$ = Dielectric loss factor.

Generally, the increased wavelength or decreased frequencies resulted in the enhanced penetration depth. The applied frequencies also greatly affect the dielectric properties of food, and there is a reduction of dielectric constant with an increase in frequency due to the reduced dipole response under high alternating frequencies (Wang et al., 2003; Franco et al., 2015). However, a smaller dielectric loss is generated on the application of a lower frequency in the aqueous solutions. However, there is a higher loss factor at lower frequencies due to the presence of free ions in solution (Heddleson and Doores, 1994).

### 2.4.2   DIELECTRIC PROPERTIES

The temperature in microwave heating is dependent on the dielectric properties of food, which are categorized as dielectric constant and dielectric loss factor, providing information about the interaction of food with the electromagnetic energy. The dielectric properties are affected by the composition of food (*MC*, salts, sugars, etc.), bulk density, temperature, and frequency used (Icier and Baysal, 2004). During microwave heating, the rate of heat generation (Q) per unit volume is quantified using the following formula:

$$Q = 2\pi\varepsilon'\varepsilon'' fE^2$$

where, E is the value of root-mean-squared (RMS) of electric field intensity (V/m).

This equation shows the increase in the rate of heat generation with increased loss factor (Sumnu and Sahin, 2005). The loss tangent also classified the food materials, and loss tangent refers to the ratio between dielectric loss and dielectric constant. The materials with higher lossy absorb higher microwave energy (Ahmed and Ramaswamy, 2007).

### 2.4.3   MOISTURE CONTENT (MC)

The *MC* in its free form and moisture evaporation rate affects the rate of heat generation during microwave heating. Increased *MC* of food results in the higher microwave absorption but reduced the penetration depth. The higher moisture foods lead to a greater dielectric loss factor, thus heating more efficiently. The decreased *MC* also increases the penetration depth but lowers the specific heat capacity to heat the product thoroughly (Heddleson and Doores, 1994; Icier and Baysal, 2004).

### 2.4.4   TEMPERATURE

As discussed earlier in section 3.2, the increased product temperature leads to water evaporation from the food, thus altering the dielectric properties. Therefore, the heat generation within the food during microwave heating also relies on the product temperature that continuously changes with moisture providing a combined effect of temperature and moisture on dielectric properties and subsequently on heating behavior (Rahman, 2007; Icier and Baysal, 2004; Heddleson and Doores, 1994).

## 2.4.5 DENSITY

The higher microwave power is absorbed by larger food samples compared to the smaller ones due to the "coupling." The mass and absorbance of microwave power exhibit a direct relationship beneficial to achieve the desired heating. The increased product mass also enhances the processing time to equilibrate the temperature gradient. The air inside food acts as an insulator decreasing the dielectric constant. Therefore, the batch oven is perfect for smaller mass objects, and continuous ovens are ideal for larger ones.

## 2.4.6 GEOMETRY AND LOCATION OF FOOD IN OVEN

The rate and uniformity of heating and penetration depth are affected by the size and shape of the product. The regular shapes result in more uniform heating due to the minimum difference in the thickness of the product. Araszkiewicz et al. (2007) exhibited the higher drying rate of a rectangular prism due to the higher surface area by microwave drying (MWD) among the different shapes of particles such as a sphere, cylinder, half-cylinder, rectangular prism, and prism with a triangle base. On the other side, the highest temperature difference was observed between the core and surface temperature of cylinder, sphere, and prism, respectively, while the minimum temperature difference was found in a prism. Generally, the particles having the largest surface area have the smallest temperatures difference between core and surface. But, Zhang and Datta (2001) revealed the more uniform heating in spherical or cylindrical shapes than square. The products having corners (90° edges) experience the more localized heating effect due to the multi-directional distribution of microwave energy. The lower penetration depth also increases the corner effects, and corner heating is less in round objects but subjected to focusing effects (Heddleson and Doores, 1994). However, the food heating in a microwave oven varies according to location also because the product at the edge of turntable is heated quickly compared to the product kept at the center (Pitachi et al., 2010).

### 2.4.7 THERMAL PROPERTIES OF FOOD

The thermal properties viz. thermal conductivity and specific heat capacity affect the microwave heating of foods. Between both, the thermal conductivity is more important to heat uniformly (Verma et al., 2017). The higher thermal conductivity makes the product to heat uniformly and rapidly than the product with lesser thermal conductivity. Specific heat capacity refers to the amount of heat gained or lost by unit product weight to temperature change of product by 1°C. The specific heat may allow a component with a poor intrinsic ability to increased temperature in the field to heat well (Heddleson and Doores, 1994). Thermal diffusivity – the function of thermal conductivity and specific heat is another property that constitutes the product's thermal properties in addition to the heat capacity and thermal conductivity (Rahman, 2007).

## 2.5 APPLICATION IN FOOD PROCESSING

In the food industry, the microwave heating has applications (Table 2.2) like cooking, dehydration, blanching, pasteurization, sterilization, tempering, and baking in addition to other applications like puffing, roasting, gelatinization, coagulation, and coating (Fu, 2010).

**TABLE 2.2**   Literature Regarding the Pasteurization of Foods Using the Microwave

| Product | Operating Conditions | System | References |
|---|---|---|---|
| Acyl gellan gel | Temperature: 22, 30, 40, 50, 60, 70, 80, 90, 100°C | | Zhang et al., 2015 |
| Apple cider | Volume: 0.5 and 1.38 lit | Continuous | Gentry and Roberts, 2005 |
| | Input power: 900, 1200, 1500, 1800, 2000 W | | |
| | Inlet temperature: 3°C, 21°C and 40°C. | | |
| | Power: 270, 450, 720, 900 W (90, 90, 60, 40 sec) | Batch | Canumir et al., 2002 |
| | Temperature: 52.5–65°C | Continuous | Tajchakavit et al., 1998 |
| | Power: 700 W | | |
| | Temperature: 73.1–96.7°C | Continuous | Villamiel et al., 1996 |

**TABLE 2.2**    *(Continued)*

| Product | Operating Conditions | System | References |
|---|---|---|---|
| Beef | Power: 2 kW–10 s Temperature: 88°C | | Tang, 2015 |
| Broccoli | Power: 3.5 kW–4.7 min | Continuous | Koskiniemi et al., 2013 |
| Carrot | Frequency: 300–3000 MHz Power: 14 kW Temperature: 22–100°C | Continuous | Peng et al., 2017 |
| Egg | Power: 1.4, 1.5, 2.5, 3.1 kW–75°C Power: 4.6, 6.4, 2.5, 2.4 kW–75°C | Batch | Zhang et al., 2014 |
| | Temperature: 0–62°C Frequency:200 MHz–10 GHz Power densities: 0.75, 1 and 2 W g⁻¹ | Batch | Dev et al., 2008 |
| Egg white | Temperature: 22–100°C Frequency: 300–3000 MHz | Batch | Zhang et al., 2013 |
| Green coconut water | Temperature: 52.5–92.9°C | — | Matsui et al., 2008 |
| | Frequency: 500–3000 MHz Temperature: 0–90°C | Continuous | Franco et al., 2015 |
| Orange juice | Flow rate:0–200 ml/min Temperature: 75–97°C | Continuous | Nikdel et al., 1993 |
| | Outlet temperature: 60°C Power: 700 W Flow rate: 137, 148, 172, 193, 240 ml/min | Continuous | Tajchakavit and Ramaswamy, 1995 |
| | Power: 700 W Exit temperature: 55, 65, 70°C Flow rate: 2–4 ml/s | Continuous | Tajchakavit and Ramaswamy, 1997a |
| | Power: 700 W Temperature: 50, 55, 60, 65°C | Batch | Tajchakavit and Ramaswamy, 1997b |
| Pickled asparagus | Temperature: 88°C Power: 2 kW–10 s; 1 kW–10 s | Batch | Lau and Tang, 2002 |
| Red bell pepper | Power: 3.5 kW–4.7 min | Continuous | Koskiniemi et al., 2013 |
| Sweet potato | Power: 3.5 kW–4.7 min | Continuous | Koskiniemi et al., 2013 |

**TABLE 2.2**     *(Continued)*

| Product | Operating Conditions | System | References |
|---|---|---|---|
| Tomato | Temperature: 22, 40, 60, 80, 100, 122°C<br>Frequency:300–3000 MHz | Batch | Peng et al., 2013 |
| Whole eggs | Temperature: 22–100°C<br>Frequency: 300–3000 MHz | Batch | Zhang et al., 2013 |

## 2.5.1  MICROWAVE HEATING IN BLANCHING

The blanching is defined as a pre-treatment carried out before the processes such as frying, freezing, drying, and canning (Xiao et al., 2017), which include the immersion of the product in hot water (HW) or steam. In this process, the time-temperature combination is essential to keep the product quality better (Reznick, 1996; Xiao et al., 2014). The normal blanching temperature is between 70°C to 100°C (Xiao et al., 2017). Generally, the better texture and retention of nutrients can be achieved at low-temperature and long-time. The temperature varying from 50–70°C (Rahman, 2007) is employed with the motive to inactivate the enzymes like peroxidase (POD), polyphenol-oxidase, lipoxygenase, and pectin enzymes which causes the off-flavors, enzymatic browning, textural changes and deterioration of nutritional quality. The common mediums used for blanching are water and steam, and their effectiveness can be judged by the inactivation of POD or polyphenol oxidase (PPO) (Rahman, 2007). As discussed above, the volumetric heating – the characteristic of microwave heating is also applied here, and heat is generated inside the food, which is different from conventional heating where the heat diffuses from outside of the food to inside. Therefore, the thermal conductivity, specific heat, and the food dimension impact the heat transfer during conventional blanching (Xin et al., 2015). The researchers revealed that microwave blanching (MWB) more effective in retaining water-soluble vitamins and other nutrients compared to conventional blanching techniques (Bingo et al., 2014; Wang, 2017). Numerous research projects has been carried out on the MWB of fruits and vegetables like carrots (Kidmose and Martens, 1999; Lemmens et al., 2009), mushroom (Devece et al., 1999), sweet potato (Liu et al., 2015), peas (Lin and Brewer, 2005), pepper (Dorantes-Alvarez et al., 2011), red bell pepper (Wang et al., 2017), and green beans (Ruiz-Ojeda and Penas, 2013).

The advantages and limitations of MWB over other blanching techniques are given in Table 2.3. Ill et al. (1998) reported the lower inactivation time of chlorophyllase of Maipo-ecotype artichokes (*Cynarascolymus* L.) using MWB than other inactivation treatments such as steam blanching (SB) and hot water blanching (HWB). The MWB also reported the complete destruction of peroxidizing with minimum loss of ascorbic acid than other methods. Wang et al. (2017) demonstrated the minimum weight loss in MWB and infrared blanching (IRB) treatments than the HWB and high humidity hot air impingement blanching (HHAIB) whereas the maximum retention of red pigments and ascorbic acid, total antioxidant activity (AA) and DPPH (2,2-diphenyl-1-picrylhydrazyl) were found in all the treatments except the HWB. The drying kinetics is thus increased by the blanching treatments.

**TABLE 2.3**   Comparison of Various Blanching Techniques

| Property | HWB | SB | MWB | HHAIB | Ohmic Blanching | IRB |
|---|---|---|---|---|---|---|
| Enzymes inactivation | √ | √ | √ | √ | √ | √ |
| Air removal from plant tissues | √ | √ | √ | √ | √ | √ |
| Reduction of microbial load | √ | √ | √ | √ | √ | √ |
| Leaching of vitamins, sugars, minerals | High | Medium | Less | Less | Less | Less |
| Water requirement | High | Medium | Less/Not required | Less | Less | Less |
| Energy requirement | High | Medium | Less | — | — | Less |
| Issues with effluent disposal | High | Moderate | Less | Less | Less | Less |
| Time used for processing | High | Higher than hot water blanching | Less | — | Less | Less |
| Blanching uniformity | Uniform heating | Non-uniform heating | Non-uniform heating | Uniform heating | — | Uniform heating |
| Weight loss | Less | Moderate | High | Less | — | High |
| Surface drying | X | √ | √ | X | √ | √ |

(*Sources*: Modified from: Bahceci et al. (2005); Lin and Brewer, (2005); Nissreen and Helen, (2006); Rungapamestry et al. (2007); Sun et al. (2007); Aguero et al. (2008); Olivera et al. (2008); Volden et al. (2009); Xin et al. (2015); Xiao et al. (2017)).

The research by Severini et al. (2016) reported the MWB as the best blanching technique to minimize the disadvantages of vegetables blanching on the color, ascorbic acid, and phenolics content. The POD using MWS is inactivated in 50 s than HWB taking 120 s and SB consuming 60 s, which showed higher vitamin C retention in MWB than other methods. The increased treatment time results in the reduction of vitamin C in both HWB and SB, which was not observed in MWB. PPO is an enzyme causing the browning reaction in mushroom, and Devece et al. (1999) revealed the complete inactivation of PPO using the combined microwave and hot-water bath treatment within a short span of time with the minimum loss of antioxidant content and increase in browning than the control. The carotene and sucrose of carrot were higher in MWB than SB and HWB (Kidmose and Martens, 1999).

Another research by Liu et al. (2015) showed the reduced drying time, quickly inactivated POD, and increased anthocyanins yield after the microwave-assisted spouted bed (MWSB) drying of purple flesh sweet potato. However, the samples treated with HWB and SB after drying were brighter and purpler than the MWB treatment, similar to steam blanched products. The microwave heating has one major drawback, i.e., the lack of uniformity in heating. The water or steam pre-treatment can be employed to increase the initial product temperature before the application of microwaves to provide an economic advantage due to the lower cost of water or steam than the microwave power. Therefore, the precise information about the power density is important to inactivate the target enzyme and to avoid undesirable effects (Shaheen et al., 2012).

### 2.5.2 MICROWAVE HEATING IN BAKING AND COOKING

The heat transfers from outside of food to inside in traditional heating, but microwave cooking generates the heat within the food by penetrating and converting into heat. The penetration depth varies from 3–5 cm, which means there is an increase in cooking time with the food thickness. The major advantage of microwave cooking is its speed, while the disadvantages are precise control of cooking time, and browning and crispness do not occur in microwave processed food (Shaheen et al., 2012). However, the combination of microwave baking with conventional can overcome the issue. The microwave-baking pan should be heat resistant and economic for industrial applications. Currently, the microwaves are employed in the baking industry for finishing when the higher baking times are consumed for low heat conductivity in the conventional process (Shaheen et al., 2012).

Paciulli et al. (2016) observed the better retention of initial texture and color quality of both raw and frozen carrots by microwave cooking than the steaming and boiling. However, the steam-cooked carrots form more soft and color than the control for both raw and frozen carrots, and the worst tissue was observed for the boiled carrots. Zhang and Hamauzu (2004) reported the heavy loss of antioxidant components and AA in broccoli during conventional and microwave cooking due to the significant difference observed in cooking. Barakat and Rohan (2014) revealed the higher kaempferol content in baked and microwaved broccoli bars than others. Polyphenols like quercetin and kaempferol are affected by frying and frying/microwaving cooking due to thermally-induced flavonoids degradation or due to their leaching into the cooking medium. On the other side, the quercetin content is increased due to the steaming and baking, while microwaving and baking increased the kaempferol levels significantly due to the disruption of plant cells and the release of bound molecules during the microwaving and steaming (Gliszczynska-Swiglo et al., 2006). The total indole GLS (glucosinolates) level were found to be decreased significantly by 53, 59, and 27% in fried, fried/microwaved, and steamed broccoli based bars, respectively, while there was no significant difference between uncooked and microwaved or baked BBBs. On the other side, the total glucosinolates content also decreased significantly during frying, frying/microwaving, steaming, and baking, except microwave cooking.

Xu et al. (2016) found the maximum retention of phenolics, flavonoids, phenolics acids, soluble sugar, anthocyanins, and AA in purple sweet potatoes (PSPs) cooked by a combination of microwave and steam than other technologies. However, the flavonol was lost significantly during the intensive microwave treatment (Rodrigues et al., 2009), while the resistant starch (RS) exhibited no significant change using the combination of microwave and steam cooking (Xu et al., 2016). Trancoso-Reyes et al. (2016) reported that the RS decreases significantly during microwaving and steaming, because of an increase in the crystallinity of sweet potato. But in conventional heating, the crystallinity destroys (Szepes et al., 2005). But, Yang et al. (2016) reported the higher retention of RS for baked and microwaved potatoes than the boiled potatoes and reduced RS in case of a combination of microwave and steam cooking than the individual microwave cooking because of steam addition and shortening the time. The suitable combination of steaming and microwave retain the active phytochemicals in PSPs and reduce the cooking time significantly than individual steaming (Xu et al., 2016).

Yang et al. (2016) reported the effect of different cooking methods like boiling, baking, and microwaving on the physical and nutritional properties of potato tubers. The application of heat during cooking significantly reduces the RS and increase in soluble starch (SS). The total starch content (TSC) was decreased during cooking. The microwave and baked potatoes report less the TSC than boiled potatoes, because of difference in the gelatinization. The reports on baking and microwave cooking revealed the increase in individual sugars like fructose, glucose, and sucrose than the raw potatoes. However, the sugars leach into the water during boiling resulting in less sugars. The total phenolic content (TPC) of tubers is affected by the cultivator genotype, and cooking techniques, and maximum TPC is retained by maintaining the reduced temperatures during boiling and higher temperatures, lower baking time and lower microwave power during microwave cooking (Yang et al., 2016). The reduction of TPC after cooking is due to the leaching into water and breakdown. The AA was increased after all the cooking methods, but the maximum was observed in microwave cooking (Yang et al., 2016). The rise in AA after cooking is due to an increase in the extractability of these compounds due to change in the structure of cellular matrix and inactivation of enzymes those are responsible for the deterioration of AA (Blessington et al., 2010; Navarre et al., 2010). The maximum weight loss was observed in baking (>20%), microwaving (<15%), and boiling (<2%), respectively (Yang et al., 2016). The weight loss during microwaving is observed due to the promoting the evaporation of cellular water (Błaszczak et al., 2004). The rate of hydrolysis of starch increases during cooking by gelatinization, so it is easily available for enzymes during digestion (Bordoloi et al., 2012). The microwave cooking reduces the availability of starch for digestive enzymes because of the formation of a crystalline structure, but the boiling destroys this structure (Mulinacci et al., 2008).

Zhong et al. (2015) observed the maximum retention of ascorbic acid in steamable microwaving (90%) and steamer steaming (90%) than the traditional microwaving (84%) of Broccoli. The cooking time for the steamable microwaving (300 s) was lower than the steamer steaming (600

s) method. The maximum antioxidant capacity of frozen broccoli was reported by steamable bag microwaving followed by steamer steaming and microwaving (53% lower than the steamable bag and 48% lower than the steamer steaming). These are due to the leaching into the cooking water during traditional microwaving. Steamable bag microwaving cooked broccoli was less soft than the traditional microwaving and steamer steaming because of the higher loss of water due to the faster release of steam allowed by its valve. The steamable bag microwaving retains the maximum color alike to uncooked broccoli, but there is an increase in $\Delta E$ value similar to traditional and steamer steaming method. Therefore, the steamable bag microwaving retains the cooking quality, color, ascorbic acid, and AA in a better way than the traditional and steamer steaming techniques (Zhong et al., 2015).

Tian et al. (2016) studied the phytochemical composition and AA of purple-fleshed potatoes as influenced by boiling, baking, steaming, microwaving, frying, stir-frying, and air frying. The microwave cooking method results in minimum loss of vitamin C compared to other cooking methods, followed by steaming, boiling, baking, stir-frying, frying, and air frying (Han et al., 2004; Tian et al., 2016). The major loss of vitamin C is during the HW boiling, frying, stir-frying, and air frying due to the leaching into water and oil and degradation during frying. But in the case of steaming and microwaving, lesser contact with water and quite low temperatures compared to frying. They also reported that microwaving results no evident change in the TPC. But the maximum retention of total anthocyanins was observed in steaming followed by microwaving. The maximum loss of total carotenoid content was observed during stir-frying, air frying, frying, and followed by microwaving. Among all cooking methods steaming and microwaving retains the maximum amount of health-promoting compounds than the stir-frying.

### 2.5.3   MICROWAVE HEATING IN DRYING

The drying is an energy-intensive process to remove moisture from a product to a pre-determined level. The energy required in the process is necessary for sensible heating and phase change of water. During the conventional drying, the product surface is highly resistant to heat transfer during the falling rate period demanding the huge amount of energy for a prolonged period to remove moisture, which results in severe product quality degradation (Feng et al., 2012). However, the combination of microwaves and conventional drying may be beneficial to overcome this problem making the process more economical. The volumetric heating and internal vapor generation is another advantage of microwave heating, leading to reduced drying time that ultimately improves product quality. The drying rate is increased by 4–8 times during MWD, and the decrease of 25–90% in drying time is compared to conventional drying (Feng et al., 2012). Here the main motto of applying microwaves is to fasten the process. The MWD can take place at atmospheric temperature or at vacuum conditions (Wang et al., 2011). The microwave heating has several advantages over conventional drying system, those are rapid heating, less processing time, energy-efficient, instantaneous, and precise control electric control, unique, and fine internal structure development, improvement of product quality and less case hardening (Feng et al., 2012; Rattanadecho and Makul, 2016). But the combination of microwave and air drying(AD) is highly beneficial where the microwave energy removes water from the product interior while the HA removes the surface water thus enhancing the drying rate and preserving the quality product (Sham et al., 2001; Andres et al., 2004; Sunjka et al., 2004). The microwave power effect was higher than the air temperature, decreasing significantly the drying time (Andres et al., 2004).

### 2.5.3.1 MICROWAVE-ASSISTED AIR DRYING (MWAD)

Numerous researches (Table 2.4) on microwave-assisted air drying (MWAD) have been found in the literature on fruits and vegetables. Nistor et al. (2017) investigated the effect of several drying techniques like free convection (50, 60, 70°C) and combination of free convection, forced convection (40°C) along with microwave power (315 W) on physical and chemical properties of red beetroot. They reported that combined drying methods were effective than the convective drying (CD) methods. The increase in drying rate and a decrease in drying time was obtained by increasing the drying temperature followed by microwave heating. The free convection (60°C), followed by forced convection (40°C) combined with microwave power 315 W is the best drying method for beetroot. Zielinska and Michalska (2016) observed the quality of blueberry as influenced by HA convective drying (HACD), MWD, and combination of HACD + MWD. The TPC and Trolox equivalent antioxidant capacity (TEAC ABTS) is reduced due to the lower air temperatures and prolonged oxygen exposure; therefore, the HACD at 60°C + MWVD led to significant degradation of TPC and TEAC ABTS values. However, there was no statistical difference in the springiness of samples dried by HACD at 90°C and HACD 90°C + MWVD from afresh sample. Thus the HACD 90°C + MWVD reported higher in product quality and low drying time out of the entire drying methods.

**TABLE 2.4** Numerous Researches Carried Out on Microwave-Assisted Air Drying (MWAD)

| Product | Temperature/Air Velocity (m/s) | Power Rate/Power (W/g) | References |
|---|---|---|---|
| Apple | 50, 60, 70°C/2 m/s | 0.1–1 W/g | Funebo et al., 2002 |
| | 50, 60, 70°C/2 m/s | — | Ahrne et al., 2003 |
| | 25, 30, 40, 50°C/1 m/s | 3, 5, 7, 10 W/g | Andres et al., 2004 |
| | 30, 50°C/2.5 m/s | 0.5 W/g | Contreras et al., 2005 |
| | 40, 60, 80°C/0.5, 1 m/s | 0.5 W/g | Funebo and Ohlsson, 1998 |
| Potato | 50, 60, 70°C/2 m/s | — | Ahrne et al., 2003 |
| | 18, 65°C/0.032 m³/s | 5, 10 W/g | Bouraoui et al., 1994 |
| | 45°C/1.5 m/s | 0.6 W/g | Jia et al., 2003 |
| Carrot | 45°C/1.5 m/s | 0.6 W/g | Jia et al., 2003 |
| | 45, 60°C/1.7 m/s | 120, 240 W | Prabhanjan et al., 1995 |
| Kiwifruit | 60°C/1.29 m/s | 210 W | Maskan, 2001 |
| Olive | 100, 160, 225°C/1 m/s | 350, 490, 700 W | Gogus and Maskan, 2001 |
| Grape | 50°C | — | Tulasidas et al., 1993 |
| Mushroom | 40, 60, 80°C/: 0.5, 1 m/s | 0.5 W/g | Funebo and Ohlsson, 1998 |
| Orange | 60°C/2 m/s | 0.17, 0.36, 0.69, 0.88 W/g | Ruiz Diaz et al., 2003 |
| Asparagus | 50, 60, 70°C | 2, 4 W/g | Nindo et al., 2003 |
| Banana | 60°C/1.45 m/s | 350, 490, 700 W | Maskan, 2000 |
| Pumpkin | 1 m/s | 160, 350 W | Alibas, 2007 |

Zielinska et al. (2015) revealed the reduction in drying time by 80% in hot air drying (HAD) + MWD than HAD on the quality of frozen/thawed and raw blueberries. The higher energy was utilized in microwave-assisted drying compared to HAD (up to 5 times) due to the great consumption of energy to create a vacuum and generate microwaves. The textural properties like hardness, chewiness, cohesiveness, and gumminess are significantly reduced in microwave-assisted drying than other drying techniques. They also reported the reduced drying time and energy consumption on initial freezing; however, the pre-treatment (freezing) increased the hardness, chewiness, and gumminess. The frozen blueberries on drying shrinks less than the ones without pre-treatment, and higher shrinkage were observed in HAD, MWVD, and HAD + MWVD thus suggesting the high-quality end product during the microwave-assisted drying of raw berries at 80°C (Zielinska et al., 2015). The drying rate is increased (minimizing the drying time) by the HA, followed by microwave finish drying (Maskan, 2000). Sharma and Prasad (2004) reported the inverse relationship between the moisture diffusivity and $MC$, which increases with microwave power and temperature at fixed air velocity. The microwave CD had reduced activation energy of garlic cloves compared to conventional drying systems (Sharma and Prasad, 2004). The higher shrinkage was found in kiwifruit dried using microwave than HAD, and less shrinkage was found in HA MWD. Thus the HA MWD demonstrated better rehydration capacity compared to other methods due to the minimum drying time (Feng and Tang, 1998; Lin et al., 1998; Maskan, 2001).

## 2.5.3.2   MICROWAVE-ASSISTED VACUUM DRYING (MWVD)

In conventional drying systems, the drying time is higher even though at higher drying temperatures due to the lower drying rates, it results in an undesirable quality dried product (Wang et al., 2011). MWVD dries the product at reduced temperature by creating a vacuum at relatively lower drying time without the absence of oxygen, resulting in the prevention of product oxidation (Table 2.5), thus maintaining the color, texture, and flavor of the product throughout the drying. The water molecules get diffused and evaporate due to the lower pressure due to the reduction of the boiling point of water, and water vapor collects at the product surface. The vacuum drying, when combined with microwave heating, improved thermal efficiency (Chandrasekaran, Ramanathan, and Basak, 2013). But the drying kinetic is slower as compared to MWVD. Usually, the quality of MWVD dried products is similar to freeze-dried products suggesting the microwave vacuum drying as a substitute for freeze-drying (FD) (Zhang, Hao, and Rui-Xin, 2010; Chandrasekaran, Ramanathan, and Tanmay, 2013).

**TABLE 2.5**  Research Work Carried Out on Microwave-Assisted Drying Vacuum Drying (MWVD)

| Product | Pressure (kPa) | Power Levels | References |
|---|---|---|---|
| Apple | 2, 4, 6.7 kPa | 425, 595, 850 kW | Kiranoudis et al., 1997 |
| Banana | 1.5, 2.5, 5, 10, 20, 30 kPa | 150, 280, 850 W | Drouzas and Schubert, 1996 |
|  | 30, 50, 101 kPa | 168 W | Mousa and Farid, 2002 |
| Beetroots | 4–6 kPa | 240, 360, 480 W | Figiel, 2010 |
| Carrot slices | 3, 5.1, 7.1 kPa | 336.5, 267.5, 162.8 W | Cui et al., 2004a |
|  | 2.5 kPa | 400, 200 W | Cui et al., 2004b |
|  | 0.3, 0.51, 0.71 kPa | 290, 359 W | Cui et al., 2005 |
|  | 13.3 kPa | 0.5, 1, 3 kW | Lin et al., 1998 |
|  | 5 kPa | 400 W | Regier et al., 2005 |
| Cranberries | 5.33, 10.67 kPa | 250, 500 W | Yongsawatdigal and Gunasekaran, 1996 |
| Garlic | 2.5 kPa | 72, 200, 400 W | Cui et al., 2005 |
|  | 4–6 kPa | 240, 480, 720 W | Figiel, 2009 |
| Garlic slices | 2.5 kPa | 400, 200, 72 W | Cui et al., 2003 |
| Grapes | 2.7 kPa | 500, 750, 1000, 1250, and 1500 W | Clary et al., 2005 |
| Kiwi | 2, 4, 6.7 kPa | 425, 595, 850 kW | Kiranoudis et al., 1997 |
| Mango | 2 kPa | 250 W | Pu and Sun, 2015–2017 |
| Mushroom | 6.5, 10, 15, 20, 23.5 kPa | 115, 150, 200, 250, 285 W | Giri and Prasad, 2007 |
| Pear | 2, 4, 6.7 kPa | 425, 595, 850 kW | Kiranoudis et al., 1997 |
| Tomatoes | 6.65 kPa | 16 kW | Durance and Wang, 2002 |

Zhao et al. (2017) reported the decrease in drying time of MWD of lotus seed due to the increased power densities (10, 15, 20 W/g) because high power densities increase the effective moisture diffusivity ($D_{eff}$)

due to the increase of kinetic energy of water molecules. On increasing the power density from 10–20 W/g, there is a decrease in drying time by 56.52%. Further, the drying time also increases the glass transmission temperature with increase in power densities and gelatinization enthalpy also increases with increasing power densities due to the annealing effect and the hydration of starch granules when the increase of microwave power density results in the samples being heated more rapidly (Bilbao-Sáinz et al., 2007; Zhong et al., 2013). They reported the application of higher microwave powers to reduce the non-enzymatic and enzymatic reactions. Therefore, the lotus seed dried at higher power levels produces a higher total free amino acid and lower starch content (Zhao et al., 2017).

The effect of MWD and MWVD on the physicochemical properties of carrot was reported by Béttega et al. (2014). The application of vacuum does not show any major changes in drying kinetics. Microwave vacuum dried carrot results in less shrinkage and more porous compared to MWD. This is advantageous; the development of porous structure increases drying rate, and it decreases drying time. β-carotene is better retained in the MWD method than MWVD at the same power levels due to the increased porosity during vacuum MWD, which led to higher permeability to vapor, thus causing higher β-carotene degradation.

Wojdyło et al. (2014) studied the influence of HACD and MWVD on drying kinetics and quality characteristics of sour cherries. An increase in air temperature during HACD and an increase in product temperature leads to deterioration of phenolic compounds (PCs), AA, and color. The better quality of cherries obtained in MWVD compared to HACD and competitive with FD (Wojdyło et al., 2014). The maximum level of AA and anthocyanin was observed during MWVD and FD when compared with AD, because of low temperature, low drying time, and lack of oxygen during drying in MWVD and FD (Leusink et al., 2010). MWVD decreases 70–90% drying time and results in better rehydration characteristics of mushroom compared with convective AD (Giri and Prasad, 2007). The convective pre-drying (CPD), in combination with MWVD, provides the high AA of beetroots similar to the FD method (Figiel, 2010). In another study, the combination of HA and vacuum MWD offers better drying than the individual (Hu et al., 2006, 2007). Generally, the drying time has an inverse relationship with the microwave power and directly related to the mass load. Further, an increase in vacuum during drying results in improved evaporation and volatilization of water from the product;

however, it may result in product overheating due to electric arcing (Hu et al., 2006; Chandrasekaran, Ramanathan, and Basak, 2013).

### 2.5.3.3   MICROWAVE-ASSISTED FREEZE DRYING (MWFD)

In freeze-drying (FD) material is frozen and then dried by sublimation process in a vacuum. The maximum quality retention of the product obtained due to a low temperature of the drying process (Genin et al., 1995; Zhang, Jiang, and Lim, 2010). Due to low operating temperature and absence of oxygen, the FD prevents chemical decomposition, oxidation of food, retention of heat-sensitive compounds in the product and good shape retention, excellent rehydration characteristics (Xu et al., 2005; Ratti, 2009; Wang et al., 2009; Jiang et al., 2010; Zhang, Jiang, and Lim, 2010). This is well suitable for heat sensitive products such as fish, meat, beef, etc. The main issues related to the conventional FD method include the higher cost, longer drying time, requirement for higher energy, and poor heat transfer rate in case of the frozen product. The FD has a 200–500% higher cost compared to HAD to achieve the same final $MC$ (Zhang, Hao, and Rui-Xin, 2010). The microwave heating may be used as a substitute for traditional conduction/radiation method and traditional conduction or radiant heating to increase the heat and mass transfer rate and the freeze-drying rate, respectively (Sunderland, 1980; Wang and Chen, 2003; Wang et al., 2009). Microwave provides the heat required for the sublimation process during drying, and the product heats volumetrically in a vacuum resulting in the increased rate of sublimation (Copson, 1958; Zhang et al., 2006, 2010; Duan et al., 2008). MWFD needs lesser drying time and energy compared to conventional FD, thus retaining the product quality, which makes it a promising technology to fasten the drying process (Wang et al., 2009). Duan et al. (2008) reported the reduction in the processing time of freeze-dried sea cucumber coated with nanosilver to half than the vacuum FD method, thus suggesting the combination of microwave freeze-drying with nanosilver coating to improve the efficiency of microbial inactivation. The sensory features and drying efficiency are not affected by a silver coating. Wang et al. (2009) summarized the effect of MWFD on drying characteristics and sensory properties of instant vegetable soup and showed an increase in drying time with an increase in thickness and a decrease in power level. The thin product layer destroys the product's sensory quality, and the better quality was obtained by using the power density of 1.0–1.5 W/g, the thickness of 15–20 mm, and temperature 50–60°C (Wang et al., 2009). The thorough study of MWFD found no commercial, industrial application due to its high costs resulting from it in a small market for freeze-dried food products (Duan et al., 2010; Chandrasekaran, Ramanathan, and Tanmay, 2013). Duan et al. (2007) reported the reduction of drying time to half than FD without any significant quality difference. The sterilization effect was exhibited by MWFD during drying, and its drying rate enhances with the increase in microwave power, decrease in cavity pressure, and material thickness. During the sublimation phase, the high microwave and cavity pressure can be used while the reverse is employed in the desorption phase, thus resulting in a quick-drying rate and better product quality (Table 2.6) (Duan et al., 2007).

**TABLE 2.6**  Research Work Carried Out on Microwave-Assisted Freeze Drying (MWFD)

| Product | Pressure | Power/Power Rate | References |
|---------|----------|------------------|------------|
| Apple | 9.15 kPa | 3.18 W/g | Li et al., 2014 |
|  | 100 Pa | 1.2, 1.4, 1.6, 1.8, 2 W/g | Wu et al., 2010 |
| Banana | 80 Pa | 1, 2, 3 W/g | Jiang et al., 2015 |
|  | 120 Pa | 2 W/g | Jiang et al., 2013 |
| Beef | 15, 65 Pa | 70 V/cm | Wang and Shi, 1999 |
| Button mushrooms | 50, 500 Pa | 1, 1.2 kW | Duan et al., 2016 |
| Cabbage | 50, 80, 100 Pa | 600, 700, 800, 900 W | Duan et al., 2007 |
| Lettuce slices | 80 Pa | 3.2 W/g | Wang et al., 2012 |

## 2.5.4  MICROWAVE HEATING IN PASTEURIZATION

Pasteurization is the process of killing key pathogens and inactivation of vegetative bacteria and enzymes present in the food. It may affect the organoleptic, nutritional, and physicochemical properties of food (Espachs-Barroso et al., 2003; Puligundla et al., 2013). The temperature and time combination is important in pasteurization to the inactive and targeted microorganism and to retain a good quality of the product. The time-temperature combination depends upon the target microorganism and nature of the product. Thermal treatment is not efficient to inactive the bacterial spores, so the product gets spoiled at room temperature.

After pasteurization immediately, the product has to be cooled to 4°C. The consumer demands safe and high quality minimally processed foods, and those traits encourage the food and academic industries for finding innovative technologies (Riahi and Ramaswamy, 2004). Due to the ability of rapid and effective heating of microwave, significant research has been started on pasteurization. Pasteurization is the process of killing pathogenic bacteria at a temperature of 60°C and 82°C. In the United States, two frequencies 915 and 2450 MHz, are designated by the Federal Communications Commission for industrial microwave heating applications. An increase in frequency leads to decrease in the penetration depth, so 915 MHz microwaves have deeper penetration depth than the 2450 MHz microwaves, and may, therefore, provide more uniform heating (Lau and Tang, 2002; Puligundla et al., 2013; Peng et al., 2017). The mechanism for the inactivation of microorganisms by microwaves is not yet clear. Some of the researchers conclude that inactivation of microorganisms is due to the generation of heat by microwaves, while other studies reported that heat may not be the only agent, although the latter group has not presented concrete evidence alternative agents. Another group reported that the microorganisms are inactivated due to the exposure of cells to microwaves effect and heating effect, it results in rupture of microbial cells (Fung and Cunningham, 1980; Ahmed and Ramaswamy, 2007). According to Kozempel et al. (1998), inactivation of microorganism takes place due to alone or a combination of the following theories:

1. **Selective Heating:** In this theory, the microorganism is heated higher than the surrounding temperature due to microwaves; this causes the destruction of microorganisms.
2. **Electroporation:** Due to the potential difference across cell membranes causes pores, it results in leakage of cellular materials.
3. **Cell Membrane Rupture:** Application of a voltage across cell membrane causes cell rupture.
4. **Magnetic Field Coupling:** The internal components of microorganisms such as DNA or protein disrupted due to electromagnetic coupling.

Villamiel et al. (1997) developed a continuous microwave pasteurization system to inactivate the microorganisms in milk. This system reports a satisfactory level of microbial inactivation without affecting the quality of the product. The use of MW heating along with other technologies like UV light, hydrogen peroxide, and γ-irradiation pronounced synergistic effects appearing to be promising food decontamination strategies. Similarly, Canumir et al. (2002) reported that MW pasteurization is also suitable to inactivate *E. coli* in apple juice. Packed acidified vegetables are pasteurized using a continuous microwave system at 915 MHz. After 60 days of storage at 30°C, there is no microbial growth has been found, and good color retention was observed, but textural properties degraded over the storage period (Koskiniemi et al., 2011). Genty and Roberts (2005) developed a continuous microwave pasteurization system for apple cider, and they evaluate the process parameters like volume load size, input power (900–2000 W), and inlet temperature (3, 21, and 40°C). They reported that the volumetric flow rate and absorbed power were criteria in the evaluation. This system consists of helical coils throughout a large cavity oven, which produces uniform heating, and the use of helical coils narrows downs the residence time. A 5 log reduction of *Escherichia coli 25922* was achieved in this study (Genty and Roberts, 2005).

The processing time for precooked carrots in MAP (microwave-assisted pasteurization) system was reduced by half for $F_{90}°C = 3$ min and 2/3 for $F_{90}°C = 10$ min when compared with HW system. The thermal deterioration of the product was less in MAP compared with the HW system. The lesser values of $\Delta E$ and higher values of a* was observed in the MAP than the HW system. There is no significant difference observed in texture and PME (pectin methylesterase) activity in both the pasteurization methods, as well as in carotenoids also (Peng et al., 2017). While using microwave heating for pasteurization, care should be taken to uniform heating of the product and avoid cold spots inside the product. The target lethal temperature is maintained for a sufficient period of time to provide a safe product (Finot, 1996; Orsat et al., 2005). Possible uneven heating, however, can be a major obstacle to the application of microwave heating in commercial pasteurization in the food industry (Wappling-Raaholt and Ohlsson, 2005). The list of research work carried out on pasteurization of foods using the microwave is illustrated in Table 2.6. A continuous flow microwave pasteurizer is a good alternative to conventional pasteurization methods. The use of microwave energy eliminates the need for a steam generator (Nikdel et al., 1993).

## 2.5.5   MICROWAVE HEATING IN STERILIZATION

Sterilization is the process of killing all microorganisms present in the food by applying high temperatures in the range of 110–130°C. The application of microwaves in the sterilization of foods is limited to the academic and industrial sectors. It is not commercialized due to the uneven heating of the product during sterilization. The temperature of the product at a few locations does not reach the real temperature distribution during microwave heating. The quality of the product during the microwave-assisted process is not better at all times. The quality degradation of food depends upon the nature of the food products, food geometry, dielectric properties, and oven designs as compared to conventional thermal processing. During microwave heating, the temperature of the product increases and the *MC* decreases these are the main two factors that affect the dielectric properties of the food, so this change in dielectric properties could the affect the heating pattern qualitatively, while such factors are not serious in conventional thermal processing (Rahman, 2007). Sun et al. (2007) reported that microwave circulated water combination (MCWC–915 MHz) treated asparagus reports higher AA and green color with higher a* values and greater hue values than the other sterilization methods (HW heating, retort heating).

## 2.5.6   MICROWAVE HEATING IN TEMPERING/THAWING

The frozen food products are tempered or thawed before it processed for further processing in the food industry. The major difference between thawing and tempering is product temperature, in thawing the temperature of a frozen product targeted to 0°C, but in case of tempering, the product is heated to below its initial freezing point (–5°C to –2°C). Tempering is preferred for the foods those are lower temperatures are acceptable for further processing (Seyhun et al., 2009; Chizoba et al., 2017; Koray and Miran, 2017). At the tempering temperature, the product firmness is intact and permits further processing without causing harm (Chizoba et al., 2017; Koray and Miran, 2017). Tempering is more commonly used in the technique for the products that are subsequently required size reduction, and it is quicker than the thawing. Tempering also reduces problems such as drip losses and bacterial growth that are associated with the thawing

(James and James, 2002). There are different conventional methods employed for thawing and tempering in food industries such as HA or water immersion methods. Heating the product with air is predominantly using the method. In these methods, heat transfers from the outside to the center of the product by conduction, it is a relatively slow process (Farag et al., 2008). Thawing at lower temperatures requires elevated processing times, however decreasing the processing time by increasing temperature leads to a decrease in product quality like drip losses and surface drying, in addition to an increase in the risk of microbial growth. Thawing with water causes the wastewater disposal problem, so another technology is required to full fill these problems (Virtanen et al., 1997). The major problems associated with these conventional methods are elevated processing time, increase in microbial load and reduction in product quality and yield (Okamoto and Suzuki, 2002; Farag et al., 2011; Oliveira et al., 2015; Uyar et al., 2015; Choi et al., 2017). Microwave thawing is a relatively new and industrially adopted technology for the thawing of food products. In microwave thawing, microwave energy penetrates into the food and produces heat internally, which leads to faster heating rates and shorter processing times and (Virtanen et al., 1997). The other advantages are low cost and improved bacterial control. The decrease in thawing time will help in a decrease in microbial growth, chemical deterioration, and drip or dehydration losses (Taher and Farid, 2001; Seyhun et al., 2009). The major problem with microwave thawing is formation localized overheating (runway heating) (Boonsumrej et al., 2007; Choi et al., 2017), it leads to major loss of water and chemical deterioration of the product, due to the absorption of microwaves by liquid water. The uniformity in heating and control of the end temperature is very important since a localized melting would be coupled to a thermal runaway effect (Swain and James, 2005). If the thawing time is less than 1 min or greater than 2000 minutes to bring down the temperature from $-8$ to $0°C$ leads to an increase in drip losses (James et al., 2002). The rate of microwave thawing depends upon the properties of the material and the magnitude and frequency of the electromagnetic radiation (Li and Sun, 2002). Li and Sun (2002) reported that different novel thawing methods and their potential application in frozen food technology like high-pressure thawing, ohmic thawing, acoustic thawing, and microwave thawing. Microwave assisted-IR heating is an efficient technique to ensure the desirable properties of food after tempering (ChizobaEkezie et al., 2017).

### 2.5.7   OTHER APPLICATION OF MICROWAVE HEATING

The other applications of microwave heating apart from the above are baking (Icoz et al., 2004; SumnuandSahin, 2005; Sumnu et al., 2007). The microwave baking required less space for power generators as compared to conventional baking systems. The accelerated baking by combining the application of MW with conventional IR baking results in enhancement of throughput, product quality, crust formation, and surface browning (OhlssonandBengtsson, 2002). Microwave baking inactivates enzymes very quickly as compared to conventional baking methods due to fast and uniform temperature rise throughout the product, to avoid starch extensive digestion and releasing sufficient $CO_2$ and stem to produce a uniform porous texture (OhlssonandBengtsson, 2002). The combined applications of microwave along with some other technologies to improve the product quality and process efficiency are MW + ultrasound (dying, extraction, and enzyme activity), MW + ohmic heating (OH) (Heating of multiphase foods) and MW + IR heating (baking, roasting, tempering, and drying) (Chizoba et al., 2017).

## 2.6   SUMMARY AND CONCLUSION

Microwave heating has advantages over conventional heating. Compared to the conventional heating systems microwave heating system shows several advantages in terms of less processing time, less cost, space-saving, and quality retention of the product. Microwaves have a wide range of applications in heating, drying, pasteurization, sterilization, and blanching, baking, cooking, thawing/tempering, and microwave-assisted extraction of compounds. The major disadvantage of microwave heating is an uneven distribution of heat. Another drawback of microwave applications in the baking industry is the lack of crust and color formation. The problems associated with conventional blanching methods such as high water requirement, leaching of nutrients, high processing time, and inefficient inactivation of enzymes. The problems with conventional drying systems are less drying rate, high drying time, and high processing time due to the casehardening, loss of quality of the product due to the elevated drying temperatures. The major problems with the conventional thawing and tempering process are microbial growth, loss of product quality, and drip losses due to high temperatures. The application of conventional methods with microwave heating is the best method to overcome these problems.

## KEYWORDS

- **dielectric constant**
- **dielectric loss factor**
- **dielectric properties**
- **dipole rotation**
- **frequency**
- **ionic polarization**
- **magnetron**
- **microwave blanching**
- **microwave-assisted drying**
- **microwave-assisted extraction**
- **microwave-assisted freeze-drying**
- **microwave-assisted pasteurization**
- **microwave-assisted sterilization**
- **microwave-assisted tempering/thawing**
- **microwave-assisted vacuum drying**

## REFERENCES

Aguero, M. V., Ansorena, M. R., Roura, S. I., & Del Valle, C. E., (2008). Thermal inactivation of peroxidase during blanching of butternut squash. *LWT Food Sci. Technol.*, *41*(3), 401–407.

Ahmed, J., & Ramaswamy, H. S., (2007). Microwave pasteurization and sterilization of foods. In: Rahman, M. S., (ed.), *Handbook of Food Preservation* (pp. 691–712). Boca Raton, FL: CRC Press.

Ahrne, L., Prothon, F., & Funebo, T., (2003).Comparison of drying kinetics and texture effects of two calcium pretreatments before microwave-assisted dehydration of apple and potato. *International Journal of Food Science and Technology, 38*, 411–420.

Alibas, I., (2007). Microwave, air and combined microwave-air-drying parameters of pumpkin slices. *LWT, 40*, 1445–1451.

Andres, A., Bilbao, C., & Fito, P., (2004). Drying kinetics of apple cylinders under combined hot air microwave dehydration. *Journal of Food Engineering, 63*, 71–78.

Araszkiewicz, M., Koziol, A., Lupinska, A., & Lupinski, M., (2007). Microwave drying of various shape particles suspended in an air stream. *Transp. Porous Med., 66*, 173–186.

Bahceci, K., Serpen, A., Gokmen, V., & Acar, J., (2005). Study of lipoxygenase and peroxidase as indicator enzymes in green beans: Change of enzyme activity, ascorbic acid and chlorophylls during frozen storage. *J. Food Eng., 66*, 187–192.

Barakat, H., & Sascha, R., (2014). Effect of different cooking methods on bioactive compounds in vegetarian, broccoli-based bars. *Journal of Functional Foods, 11*, 407–416.

Benlloch-Tinoco, M., Igual, M., Salvador, A., Rodrigo, D., & Martínez-Navarrete, N., (2014). Quality and acceptability of microwave and conventionally pasteurized kiwifruit puree. *Food Bioprocess Technol., 7*(11), 3282–3292. http://dx.doi.org/10.1007/s11947-014-1315-9 (Accessed on 8 November 2019).

Benlloch-Tinoco, M., Marta, I., Dolores, R., & Nuria, M. N., (2013). Comparison of microwaves and conventional thermal treatment on enzymes activity and antioxidant capacity of kiwifruit puree.*Innovative Food Science and Emerging Technologies, 19*, 166–172.

Béttega, R., Rosa, J. G., Corrêa, R. G., &Freire, J. T., (2014). Comparison of carrot (Daucuscarota) drying in microwave and in vacuum microwave. *Brazilian Journal of Chemical Engineering, 31*(02), pp. 403–412.

Bilbao-Sáinz, C., Butler, M., Weaver, T., & Bent, J., (2007). Wheat starch gelatinization under microwave irradiation and conduction heating.*Carbohydrate Polymers, 69*(2), 224–232.

Bilecka, I., & Niederberger, M., (2010). Microwave chemistry for inorganic nanomaterials synthesis. *Nanoscale, 2*, 1358–1374.

Bingol, G., Wang, B., Zhang, A., Pan, Z., & McHugh, T. H., (2014). Comparison of water and infrared blanching methods for processing performance and final product quality of French fries. *Journal of Food Engineering, 121*, 135–142.

Błaszczak, W., Sadowska, J., Fornal, J., Vacek, J., Flis, B., & Zagórski-Ostoja, W., (2004). Influence of cooking and microwave heating on microstructure and mechanical properties of transgenic potatoes. *Nahrung/Food, 48*(3), 169–176.

Blessington, T., Nzaramba, M. N., Scheuring, D. C., Hale, A. L., Reddivari, L., & Miller, J. C., (2010). Cooking methods and storage treatments of potato: Effects on carotenoids, antioxidant activity, and phenolics. *American of Journal of Potato Research, 87*(6), 479–491.

Boonsumrej, S., Chaiwanichsiri, S., Tantratian, S., Suzuki, T., & Takai, R., (2007). Effects of freezing and thawing on the quality changes of tiger shrimp (Penaeusmonodon) frozen by air-blast and cryogenic freezing. *Journal of Food Engineering, 80*, 292–299.

Bordoloi, A., Kaur, L., & Singh, J., (2012). Parenchyma cell microstructure and textural characteristics of raw and cooked potatoes. *Food Chemistry, 133*(4), 1092–1100.

Bouraoui, M., Richard, P., & Durance, T., (1994). Microwave and convective drying of potato slices.*Journal of Food Process Engineering, 17*, 353–363.

Canumir, J. A., Celis, J. E., De Bruijn, J., & Vidal, L. V., (2002). Pasteurization of apple juice by using microwaves. *Lebensmittel-Wissenschaft und-Technologie, 35*, 389–392.

Chandrasekaran, S., Ramanathan, S., & Basak, T., (2013). Microwave food processing — a review. *Food Research International, 52*(1), 243–261.

Chandrasekaran, S., Ramanathan, S., &Tanmay, B., (2013).Microwave food processing—a review.*Food Research International, 52*(1), 243–261.

Chizoba, E. F. G., Sun, D. W., Han, Z., & Cheng, J. H., (2017). Microwave-assisted food processing technologies for enhancing product quality and process efficiency: A review of recent developments. *Trends in Food Science and Technology,67,* 58–69.

Choi, E. J., Hae, W. P., Young, B. C. A., Sung, H. P. B., Jin, S. K., & Ho-Hyun, C., (2017). Effect of tempering methods on quality changes of pork loin frozen by cryogenic immersion.*Meat Science, 124,* 69–76.

Cinquanta, L., Di Matteo, M., Cuccurullo, G., & Albanese, D., (2010).Effect on orange juice of batch pasteurization in an improved pilot-scale microwave oven.*Journal of FoodScience, 75,* 46–50.

Clary, C. D., Wang, S. J., &Petrucci, V. E., (2005). Fixed and incremental levels of microwave power application on drying grapes under vacuum. *Journal of Food Science,70*(5), 344–349.

Contreras, C., Martin, M. E., Martinez-Navarrete, N., &Chiralt, A., (2005). Effect of vacuum impregnation and microwave application on structural changes which occurred during air-drying of apple. *LWT-Food Science and Technology, 38,* 471–477.

Copson, D. A., (1958). Microwave sublimation of foods. *Food Technology, 12,* 270–272.

Cui, Z. W., Xu, S. Y., & Sun, D. W., (2003). Dehydration of garlic slices by combined microwave-vacuum and air drying. *Drying Technology, 21*(7), 1173–1184.

Cui, Z. W., Xu, S. Y., & Sun, D. W., (2004a). Microwave-vacuum drying kinetics of carrot slices. *Journal of Food Engineering, 65*(2), 157–164.

Cui, Z. W., Xu, S. Y., & Sun, D. W., (2004b). Effect of microwave-vacuum drying on the carotenoids retention of carrot slices and chlorophyll retention of Chinese chive leaves. *Drying Technology, 22*(3), 561–574.

Cui, Z. W., Xu, S. Y., Sun, D. W., & Chen, W., (2005). Temperature changes during microwave-vacuum drying of sliced carrots. *Drying Technology, 23*(5), 1057–1074.

Datta, A. K., Sumnu, G., &Raghavan, G., (2005). Dielectric properties of food. In: Rao, M. A., Rizvi, S., &Datta, A. K., (eds.), *Engineering Properties of Foods.* CRC Press, Boca Raton.

Dev, S. R. S., Gariepy, Y., &Raghavan, G. S. V., (2008). Dielectric properties of egg components and microwave heating for in-shell pasteurization of eggs.*Journal of FoodEngineering, 86,* 207–214.

Devece, C., Jose, N. R. L., Lorena, G. F., Jose, T., Jose, M. C., Elias, D. L. R., & Francisco, G. C., (1999). Enzyme inactivation analysis for industrial blanching applications: Comparison of microwave, conventional, and combination heat treatments on mushroom polyphenoloxidase activity. *J. Agric. Food Chem., 47,* 4506–4511.

Dorantes-Alvarez, L., Jaramillo-Flores, E., Gonzalez, K., Martinez, R., &Parada, L., (2011). Blanching peppers using microwaves. *Procedia Food Science, 1,* 178–183.

Drouzas, A. E., & Schubert, H., (1996). Microwave application in vacuum drying of fruits. *Journal of Food Engineering, 28,* 203–209.

Duan, X., Liu, W. C., Ren, G. Y., Liu, L. L., & Liu, Y. H., (2016). Browning behavior of button mushrooms during microwave freeze-drying. *Drying Technology, 34*(11), 1373–1379.

Duan, X., Min, Z., &Arun, S. M., (2007). Studies on the microwave freeze drying technique and sterilization characteristics of cabbage. *Drying Technology, 25*, 1725–1731.

Duan, X., Min, Z., Xinlin, L., &Arun, S. M., (2008). Microwave freeze drying of sea cucumber coated with nanoscale silver. *Drying Technology, 26*, 413–419.

Duan, X., Zhang, M., & Mujumdar, A. S., (2007). Studies on the microwave freeze drying technique and sterilization characteristics of cabbage. *Drying Technology, 10*(25), 1725–1731.

Duan, X., Zhang, M., Mujumdar, A. S., & Wang, R., (2010). Trends in microwave assisted freeze drying of foods. *Drying Technology, 28*(4), 444–453.

Durance, T. D., & Wang, J. H., (2002). Energy consumption, density and rehydration rate of vacuum microwave- and hot-air convection-dehydrated tomatoes. *Journal of Food Science,67*, 2212–2216.

Espachs-Barroso, A., Barbosa-Canovas, G. V., & Martin-Belloso, O., (2003). Microbial and enzymatic changes in fruit juice induced by high-intensity pulsed electric fields. *Food Rev. Int., 19*, 253–273.

Farag, K. W., Cronin, D. A., Morgan, D. J., & Lyng, J. G., (2008). Dielectric and thermophysical properties of different beef meat blends over a temperature range of $-18$ to $+10°C$. *Meat Science, 79*, 740–747.

Farag, K. W., Lyng, J. G., Morgan, D. J., & Cronin, D. A., (2008). A comparison of conventional and radio frequency tempering of beef meats: Effects on product temperature distribution. *Meat Science, 80*, 488–495.

Farag, K. W., Lyng, J. G., Morgan, D. J., & Cronin, D. A., (2011). A comparison of conventional and radio frequency thawing of beef meats: Effects on product temperature distribution. *Food Bioprocess Technol., 4*, 1128–1136.

Feng, H., & Tang, J., (1998). Microwave finish drying of diced apples in a spouted bed. *Journal of Food Science, 63*, 679–683.

Feng, H., Tang, J., & Cavalieri, R. P., (2002). Dielectric properties of dehydrated apples as affected by moisture and temperature. *Transaction of ASAE, 45*(1), 129–135.

Feng, H., Yun, Y., & Juming, T. (2012). Microwave drying of food and agricultural materials: Basics and heat and mass transfer modeling. *Food Eng Rev.*doi: 10.1007/s12393-012-9048-x.

Figiel, A., (2009). Drying kinetics and quality of vacuum-microwave dehydrated garlic cloves and slices. *Journal of Food Engineering, 94*, 98–104.

Figiel, A., (2010). Drying kinetics and quality of beetroots dehydrated by combination of convective and vacuum-microwave methods. *Journal of Food Engineering, 98*, 461–470.

Finot, P. A., (1996). Effect of microwave treatments on the nutritional quality of foods. *Cahiers de Nutrition et de Dietetique, 31*(4), 239–246.

Franco, A. P., Letícia, Y., Yamamoto, C. C., Tadini, J. A., & Gut, W., (2015). Dielectric properties of green coconut water relevant to microwave processing: Effect of temperature and field frequency. *Journal of Food Engineering, 155*, 69–78.

Fu, Y. C., (2010). Microwave heating in food processing. In: Hui, Y. H., (ed.), *Handbook of Food Science, Technology and Engineering* (pp. 1–11). Boca Raton, FL: CRC Press.

Funebo, T., & Ohlsson, T., (1998). Microwave-assisted air dehydration of apple and mushroom. *Journal of Food Engineering, 38*, 353–367.

Funebo, T., Ahrne, L., Prothon, F., Kidman, S., Langton, M., & Skjoldebrand, C., (2002). Microwave and convective dehydration of ethanol treated and frozen apple-physical properties and drying kinetics. *International Journal of Food Science and Technology, 37*, 603–614.

Fung, D., & Cunningham, F., (1980). Effect of microwaves on microorganisms in foods. *Journal of Food Protection, 43*(8), 641–650.

Genin, N., & Rene, F., (1995). Analysis of the role of the glass transition in the methods of food preservation. *Journal of Food Engineering, 26*, 391–408.

Gentry, T. S., & Roberts, J. S., (2005). Design and evaluation of a continuous flow microwave pasteurization system for apple cider. *LWT Food Science and Technology, 38*, 227–238.

Giri, S. K., Suresh, P., (2007). Drying kinetics and rehydration characteristics of microwave-vacuum and convective hot-air dried mushrooms. *Journal of Food Engineering, 78*, 512–521.

Gliszczynska-Swiglo, A., Ciska, E., Pawlak-Lemanska, K., Chmielewski, J., Borkowski, T., & Tyrakowska, B., (2006). Changes in the content of health-promoting compounds and antioxidant activity of broccoli after domestic processing. *Food Additives andContamination, 23*, 1088–1098.

Gogus, F., & Maskan, M., (2001). Drying of olive pomace by a combined microwave-fan assisted convection oven. *Nahrung, 45*(2), 129–132.

Guan, D., Cheng, M., Wang, Y., & Tang, J., (2004). Dielectric properties of mashed potatoes relevant to microwave and radiofrequency pasteurization and sterilization processes. *Journal of Food Science, 69*, 30–37.

Guo, Q., Da-Wen, S., Jun-Hu, C., & Zhong, H., (2017). Microwave processing techniques and their recent applications in the food industry. *Trends in Food Science and Technology, 67*, 236–247.

Han, J., Kozukue, N., Young, K., Lee, K., & Friedman, M., (2004). Distribution of ascorbic acid in potato tubers and in home-processed and commercial potato foods. *Journal of Agricultural and Food Chemistry, 52*(21), 6516–6521.

Heddleson, R. A., & Doores, S. (1994). Factors affecting microwave heating of foods and microwave induced destruction of foodborne pathogens—a review. *Journal of Food Protection, 57*(11), 1025–1037.

Hu, Q. G., Zhang, M., Mujumdar, A. S., Xiao, G. N., & Sun, J. C., (2006). Drying of edamames by hot air and vacuum microwave combination. *Journal of Food Engineering, 77*, 977–982.

Hu, Q. G., Zhang, M., Mujumdar, A. S., Xiao, G. N., & Sun, J. C., (2007). Performance evaluation of vacuum microwave drying of edamame in deep-bed drying. *Drying Technology, 25*, 731–736.

Icier, F., & Baysal, T., (2004). Dielectrical properties of food materials—1: Factors affecting and industrial uses. *Critical Reviews in Food Science and Nutrition, 44*, 465–471.

Icoz, D., Sahin, S., & Sumnu, G., (2004). Color and texture development during microwave and conventional baking of breads. *International Journal of Food Properties, 7*, 201–213.

Ihl, M., Marta, M., & Valerio, B., (1998). Chlorophyllase inactivation as a measure of blanching efficacy and color retention of artichokes (*Cynarascolymus* L.). *Lebensm.-Wiss. u.-Technol., 31*, 50–56.

James, C., Swain, M. V. L., James, S. J., & Swain, M. J., (2002). Development of a methodology for assessing the heating performance of a domestic microwave oven. *International Journal of Food Science and Technology, 37*(8), 879–892.

James, S. J., & James, C., (2002). Thawing and tempering. In: *Meat Refrigeration* (pp. 159–190). Woodhead Publishing, Cambridge.

Jia, L. W., Islam, M. R., & Mujumdar, A. S., (2003). A simulation study on convection and microwave drying of different food products. *Drying Technology, 21*(8), 1549–1574.

Jiang, H., Zhang, M., & Mujumdar, A. S., (2010). *Microwave Freeze-Drying Characteristics of Banana Crisps, 28*(12), 1377–1384.

Jiang, H., Zhang, M., Mujumdar, A. S., & Lim, R. X., (2013). Analysis of temperature distribution and SEM images of microwave freeze drying banana chips. *Food and Bioprocess Technology, 6*(5), 1144–1152.

Jiang, H., Zhang, M., Mujumdar, A. S., & Lim, R. X., (2015). Drying uniformity analysis of pulse-spouted microwave-freeze drying of banana cubes. *Drying Technology, 34*(5), 539–546.

Kidmose, U., & Kaack, K., (1999). Changes in texture and nutritional quality of green asparagus spears (*Asparagus officinalis* L.) during microwave blanching and cryogenic freezing. *ActaAgriculturaeScandinavica, Section B-Soil and Plant Science, 49*, 110–116.

Kidmose, U., & Martens, H. J., (1999). Changes in texture, microstructure and nutritional quality of carrot slices during blanching and freezing.*Journal of the Science of Food and Agriculture, 79*, 1747–1753.

Kiranoudis, C. T., Tsami, E., & Maroulis, Z. B., (1997). Microwave vacuum drying kinetics of some fruits. *Drying Technology, 15*, 2421–2440.

Koray, P. T., & Welat, M., (2017). Experimental comparison of microwave and radio frequency tempering of frozen block of shrimp. *Innovative Food Science and Emerging Technologies, 41*, 292–300.

Koskiniemi, C. B., McFeeters, R. F., Simunovic, J., & Truong, V. D., (2011). Improvement of heating uniformity in packaged acidified vegetables pasteurized with a 915MHz continuous microwave system. *Journal of Food Engineering, 105*, 149–160.

Koskiniemi, C. B., Truong, V. D., Mcfeeters, R. F., & Simunovic, J., (2013). Quality evaluation of packaged acidified vegetables subjected to continuous microwave pasteurization. *LWT-Food Sci. Technol., 54*, 157–164.

Kozempel, M. F., Annous, B. A., Richard, D., CooK, O. J. S., & Rchard, C. W., (1998). Inactivation of microorganisms with microwaves at reduced temperatures. *Journal of Food Protection, 61*(5), 582–585.

Krishnamoorthy, P., Sohan, L. B., Jeyamkondan, S., & David, D. J., (2010). Heating performance assessment of domestic microwave ovens. *International Microwave Power Institute's 44th Annual Symposium DENVER 44th Symposium.*

Lau, M. H., & Tang, J., (2002). Pasteurization of pickled asparagus using 915 MHz microwaves. *Journal of Food Engineering, 51*, 283–290.

Lemmens, L., Evelina, T., Cecilia, S., Chantal, S., Lília, A., Maud, L., Marie, A., Ann, V. L., & Marc, H., (2009). Thermal pretreatments of carrot pieces using different heating techniques: Effect on quality related aspects. *Innovative Food Science and Emerging Technologies, 10*, 522–529.

Leusink, G. J., David, D. K., Yaghmaee, P., & Durance, T., (2010). Retention of antioxidant capacity of vacuum microwave dried cranberry. *Journal of Food Science, 75*(3), C311–C316.

Li, B., & Sun, D. W., (2002). Novel methods for rapid freezing and thawing of foods – a review. *Journal of Food Engineering, 54*, 175–182.

Li, R., Huang, L., Zhang, M., Mujumdar, A. S., & Wang, Y. C., (2014). Freeze drying of apple slices with and without application of microwaves. *Drying Technology, 32*(15), 1769–1776.

Lin, S., & Brewer, M. S., (2005). Effects of blanching method on the quality characteristics of frozen peas. *J. Food Qual., 28*, 350–360.

Lin, T. M., Durance, T. D., & Scaman, C. H., (1998). Characterization of vacuum microwave, air and freeze dried carrot slices. *Food Research International, 4*, 111–117.

Liu, P., Arun, S., Mujumdar, M. Z., &Hao, J., (2015). Comparison of three blanching treatments on the color and anthocyanin level of the microwave-assisted spouted bed drying of purple flesh sweet potato. *Drying Technology, 33*, 66–71.

Maria, P. M., Tommaso, G., Eleonora, C., Nicoletta, P., Alessandro, P., & Emma, C., (2016). Effect of different cooking methods on structure and quality of industrially frozen carrots. *Food Sci. Technol., 53*(5), 2443–2451.

Maskan, M., (2000). Microwave/air and microwave finish drying of banana. *Journal of Food Engineering, 44*, 71–78.

Maskan, M., (2001). Drying, shrinkage and rehydration characteristics of kiwifruits during hot air and microwave drying. *Journal of Food Engineering, 48*, 177–182.

Maskan, M., (2001). Kinetics of color change of kiwifruits during hot air and microwave drying. *Journal of Food Engineering, 48*, 169–175.

Matsui, K. N., Gut, J. A. W., De Oliveira, P. V., & Tadini, C. C., (2008). Inactivation kinetics of polyphenol oxidase and peroxidase in green coconut water by microwave processing. *J. Food Eng., 88*, 169–176. http://dx.doi.org/10.1016/j.jfoodeng.2008.02.003 (Accessed on 8 November 2019).

Metaxas, A. C., & Meredith, R. J., (1983).Industrial microwave heating. *IEE Power Engineering Series*. London: Peregrinus.

Mousa, N., & Farid, M., (2002). Microwave vacuum drying of banana slices. *Drying Technology, 20*, 2055–2066.

Mulinacci,N.,Leri,F.,Giaccherini,C.,Innocenti,M.,Andrenelli,L.,Canova,G.,&Casiraghi, M. C., (2008). Effect of cooking on the anthocyanins, phenolic acids, glycoalkaloids, and resistant starch content in two pigmented cultivars of *Solanumtuberosum* L. *Journal of Agricultural and Food Chemistry, 56*(24), 11830–11837.

Narang, S., & Gupta, M., (2015). Energy harvesting from space based solar power satellite. *International Journal of Advance Research in Science and Engineering, 4*(1), 11–15.

Navarre, D. A., Shakya, R., Holden, J., & Kumar, S., (2010). The effect of different cooking methods on phenolics and vitamin C in developmentally young potato tubers.*American Journal of Potato Research, 87*, 350–359.

Nikde, S., Chin, S., Chen, J., Mickey, E. P., Donald, G. M., & Lorrie, M. F., (1993). Pasteurization of citrus juice with microwave energy in a continuous-flow unit. *J. Agdc. Food Chem., 41*, 2116–2119.

Nindo, C., Ting, S., Wang, S. W., Tang, J., & Powers, J. R., (2003). Evaluation of drying technologies for retention of physical and chemical quality of green asparagus (*Asparagus officinalis* L.). *LWT-Food Science and Technology, 36*(5), 507–516.

Nissreen, A. G., & Helen, C., (2006). The effect of low temperature blanching on the texture of whole processed new potatoes. *J. Food Eng., 74*, 335–344.

Nistor, O. V., Liliana, S. C., Doina, G. A., Ludmila, R., & Elisabeta, B., (2017). Influence of different drying methods on the physicochemical properties of red beetroot (*Beta vulgaris* L. var. Cylindra).*Food Chemistry, 236*, 59–67.

Ohlsson, T., & Bengtsson, N., (2002). Minimal processing of foods with nonthermal methods. In: Ohlsson, T., &Bengtsson, N., (eds.), *Minimal Processing Technologies in the Food Industry* (pp. 34–60). Cambridge: Woodhead Publishing.

Okamoto, A., & Suzuki, A., (2002). Effects of high hydrostatic pressure thawing on pork meat. In: Hayashi, R., (ed.), *Trends in High Pressure Bioscience and Biotechnology (Progress in Biotechnology)* (Vol. 19, pp. 571–576). Amsterdam, The Netherlands: Elsevier Science B.V.

Okeke, C. Abioye, A. E., & Omosun, Y., (2014). Microwave heating applications in food processing. *IOSR Journal of Electrical and Electronics Engineering (IOSR-JEEE), 9*(4-II), 29–34.

Oliveira, M. R., Gubert, G., Roman, S. S., Kempka, A. P., & Prestes, R. C., (2015). Meat quality of chicken breast subjected to different thawing methods. *Brazilian Journal of Poultry Science, 17*, 165–172.

Olivera, D. F., Vina, S. Z., Marani, C. M., Ferreyra, R. M., Mugridge, A., Chaves, A. R., & Mascheroni, R. H., (2008). Effect of blanching on the quality of Brussels sprouts (*Brassica oleracea* L. gemmifera DC) after frozen storage.*J. Food Eng., 84*, 148–155.

Orsat, V., Raghavan, V., & Meda, V., (2005). Microwave technology for food processing: An overview. In: Schubert, H., & Regier, M., (eds.), *The Microwave Processing of Foods* (pp. 105–115). Cambridge: Woodhead.

Peng, J., Tang, J., Donglei, L., Frank, L. B., Zhongwei, T. B., Feng, L. B. C., & Wenjia, Z., (2017). Microwave pasteurization of pre-packaged carrots. *Journal of Food Engineering, 202*, 56–64.

Peng, J., Tang, J., Yang, J., Stewart, G. Bohnet, D., & Barrett, M., (2013). Dielectric properties of tomatoes assisting in the development of microwave pasteurization and sterilization processes. *LWT – Food Science and Technology, 54*, 367–376.

Prabhanjan, D. G., Ramaswamy, H. S., & Raghavan, G., (1995). Microwave-assisted convective air drying of thin layer carrots. *Journal of Food Engineering, 25*, 283–293.

Pu, Y. Y., & Sun, D. W., (2015). Vis-NIR hyperspectral imaging in visualizing moisture distribution of mango slices during microwave-vacuum drying. *Food Chemistry, 188*, 271–278.

Pu, Y. Y., & Sun, D. W., (2016). Prediction of moisture content uniformity of microwave-vacuum dried mangoes as affected by different shapes using NIR hyperspectral imaging. *Innovative Food Science and Emerging Technologies, 34*, 348–356.

Pu, Y. Y., & Sun, D. W., (2017). Combined hot-air and microwave-vacuum drying for improving drying uniformity of mango slices based on hyperspectral imaging visualization of moisture content distribution. *Biosystems Engineering, 156*, 108–119.

Puligundla, P., Seerwan, A. A., Won, C., Soojin, J., Sang-Eun, O., & Sanghoon, K., (2013). Potentials of microwave heating technology for select food processing applications – a brief overview and update. *Food Process Technol., 4*, 11.

Rahman, M. S., (2007). *Handbook of Food Preservation* (2nd edn.). New York: CRC Press.

Rattanadecho, P., & Makul, N., (2016). Microwave-assisted drying: A review of the state-of-the-art. *Drying Technology, 34*(1), 1–38.

Ratti, C., (2001). Hot air and freeze-drying of high-value foods: A review. *Journal of Food Engineering, 49*(4), 311–319.

Regier, M., Mayer-Miebach, E., Behsnilian, D., Neff, E., & Schuchmann, A., (2005). Influences of drying and storage of lycopene-rich carrots on the carotenoid content. *Drying Technology, 23*, 989–998.

Regier, M., Rother, M., & Schuchmann, H. P., (2010). Alternative heating technologies. In: Ortega-Rivas, E., (ed.), *Processing Effects on Safety and Quality of Foods* (pp. 188–245). Boca Raton, FL: CRC Press.

Reznick, D., (1996). Ohmic heating of fluid foods: Ohmic heating for thermal processing of foods: government, industry, and academic perspectives. *Food Technol., 50*, 250, 251.

Riahi, E., & Ramaswamy, H. S., (2004). High pressure inactivation kinetics of amylase in apple juice. *J. Food Eng., 64*, 151–160.

Rodrigues, A. S., Pérez-Gregorio, M. R., García-Falcón, M. S., & Simal-Gándara, J., (2009). Effect of curing and cooking on flavonols and anthocyanins in traditional varieties of onion bulbs. *Food Research International, 42*(9), 1331–1336.

Ronald, A. H., & Stephanie, D., (1994). Factors affecting microwave heating of foods and microwave induced destruction of foodborne pathogens – A review. *Journal of Food Protection, 57*(11), 1025–1037.

Ruiz, D. G., Martınez-Monzo, J., Fito, P., & Chiralt, A., (2003). Modeling of dehydration rehydration of orange slices in combined microwave/air drying. *Innovative Food Science and Emerging Technologies, 4*(2), 203–209.

Ruiz-Ojeda, L. M., & Penas, F. J., (2013). Comparison study of conventional hot-water and microwave blanching on quality of green beans. *Innovative Food Science and Emerging Technologies, 20*, 191–197.

Rungapamestry, V., Duncan, A. J., Fuller, Z., & Ratcliffe, B., (2007). Effect of cooking Brassica vegetables on the subsequent hydrolysis and metabolic fate of glucosinolates. *P. Nutr. Soc., 66*, 69–81.

Severini, C., Giuliani, R., De Filippis, A., Derossil, A., & De Pilli, T., (2016). Influence of different blanching methods on color, ascorbic acid and phenolics content of broccoli. *J. Food Sci. Technol., 53*(1), 501–510.

Seyhun, N., Sahin, S., Ahmed, J., & Sumnu, G., (2009). Comparison and modeling of microwave tempering and infrared assisted microwave tempering of frozen potato puree. *Journal of Food Engineering, 92*, 339–344.

Shaheen, M. S., Khaled, F. E. M., Ahmed, H. E. G., & Faqir, M. A. Microwave applications in thermal food processing. In: Wenbin, C., (ed.), *The Development and Application of Microwave Heating* (pp. 3–16).

Sham, P. W. Y., Scaman, C. H., & Durance, T. D., (2001). Texture of vacuum microwave dehydrated apple chips as affected by calcium pretreatment, vacuum level, and apple variety. *Journal of Food Science, 66*(9), 1341–1347.

Sharma, G. P., & Prasad, S., (2004). Effective moisture diffusivity of garlic cloves undergoing microwave-convective drying. *Journal of Food Engineering, 65,* 609–617.

Sosa-Morales, M. E., Valerio-Junco, L., Lopez-Malo, A., & Garcia, H. S., (2010). Dielectric properties of food: Reported data in the 21st century and their potential applications. *Food Sci. Technol., 43,* 1169–1179. http://dx.doi.org/10.1016/j.lwt.2010.03.017 (Accessed pm 8 November 2019).

Sumnu, G., & Sahin, S., (2005). Recent developments in microwave heating. In: Sun, D. W., (ed.), *Emerging Technologies for Food Processing* (pp. 419–444). California: Elsevier Academic Press.

Sumnu, G., Keskin, S. O., Rakesh, V., Datta, A. K., & Sahin, S., (2007). Transport and related properties of breads baked using various heating modes. *Journal of Food Engineering, 78,* 1382–1387.

Sumnu, S. G., & Sahin, S. (2012). Microwave heating. In: Da-Wen, S., (ed.), *Thermal Food Processing: New Technologies and Quality Issues* (pp. 555–576).

Sun, T., Tang, J. M., & Joseph, R., (2007). Powers antioxidant activity and quality of asparagus affected by microwave-circulated water combination and conventional sterilization. *Food Chem., 100,* 813–819.

Sunderland, J. E., (1980). Microwave freeze drying. *Journal of Food Process Engineering, 4*(4), 195–212.

Sunjka, P. S., Rennie, T. J., Beaudry, C., & Raghavan, G. S. V., (2004). Microwave–convective and microwave–vacuum drying of cranberries: A comparative study. *Drying Technology, 22*(5), 1217–1231.

Swain, M., & James, S., (2005). Microwave thawing and tempering. In: Schubert, H., & Regier, M., (eds.), *The Microwave Processing of Foods* (pp. 174–190). Cambridge: Woodhead.

Szepes, A., Hasznos-Nezdei, M., Kovacs, J., Funke, Z., Ulrich, J., & Szabo-Revesz, P., (2005). Microwave processing of natural biopolymers–studies on the properties of different starches.*International Journal of Pharmaceutics, 302*(1&2), 166–171.

Taher, B. J., & Farid, M. M., (2001). Cyclic microwave thawing of frozen meat: Experimental and theoretical investigation. *Chemical Engineering and Processing, 40*(4), 379–389.

Tajchakavit, S., & Ramaswamy, H. S., (1995). Continuous-flow microwave heating of orange juice: Evidence of nonthermal effects. *Journal of Microwave Power and Electromagnetic Energy, 30,* 141–148.

Tajchakavit, S., & Ramaswamy, H. S., (1997a). Continuous-flow microwave inactivation kinetics of pectin methyl esterase in orange juice.*Journal of Food Processing and Preservation, 21,* 365–378.

Tajchakavit, S., & Ramaswamy, H. S., (1997b). Thermal vs. Microwave inactivation kinetics of pectin methylesterase in orange juice under batch mode heating conditions. *Lebensmittel-Wissenschaft und-Technologie, 30,* 85–93.

Tajchakavit, S., Ramaswamy, H. S., & Fustier, P., (1998). Enhanced destruction of spoilage microorganisms in apple juice during continuous flow microwave heating. *Food Research International, 31,* 713–722.

Tang, J., (2015). Unlocking potentials of microwaves for food safety and quality. *J. Food Sci., 80,* E1776–E1793.

Tewari, G., (2007). Microwave and radiofrequency heating. In: Tewari, G., & Juneja, V. K., (eds.), *Advances in Thermal and Non- Thermal Food Preservation* (pp. 91–98). Ames, IA: Blackwell Publishing.

Tian, J., Jianle, C., Feiyan, L., Shiguo, C., Jianchu, C., Donghong, L., & Xingqian, Y., (2016). Domestic cooking methods affect the phytochemical composition and antioxidant activity of purple-fleshed potatoes. *Food Chemistry, 197*, 1264–1270.

Trancoso-Reyes, N., Ochoa-Martinez, L. A., Bello-Perez, L. A., Morales-Castro, J., Estevez-Santiago, R., & Olmedilla-Alonso, B., (2016). Effect of pre-treatment on physicochemical and structural properties, and the bioaccessibility of beta-carotene in sweet potato flour. *Food Chemistry, 200*, 199–205.

Tulasidas, T. N., Raghavan, G. S. V., & Norris, E. R., (1993). Microwave and convective drying of grapes.*Transactions of the ASAE, 36*(6), 1861–1865.

Uyar, R., Bedane, T. F., Erdogdu, F., Palazoglu, T. K., Farag, K. W., & Marra, F., (2015). Radiofrequency thawing of food products—a computational study. *Journal of Food Engineering, 146*, 163–171.

Venkatesh, M. S., & Raghavan, G. S. V., (2004). An overview of microwave processing and dielectric properties of agri-food materials. *Biosystems Engineering, 88*(1), 1–18.

Verma, D. K., Kapri, M., Billoria, S., Mahato, D. K., & Prem, P. S., (2017). Effects of thermal processing on nutritional composition of green leafy vegetables: A review. In: Verma, D. K., & Goyal, M. R., (eds.), *Engineering Interventions in Foods and Plants*. As part of book series on "Innovations in Agricultural and Biological Engineering",Apple Academic Press, USA.

Villamiel, M., Lopez-Fandino, R., & Olano, A., (1997). Microwave pasteurization of milk in a continuous flow unit: Effects on the cheese-making properties of goat's milk. *Milchwissenschaft, 52*, 29–32.

Villamiel, M., Lopez-Fandino, R., Corzo, N., Martinez-Castro, I., & Olano, A., (1996). Effects of continuous-flow microwave treatment on chemical and microbiological characteristics of milk. *Zeitsrift f. urLebensmittelUntersuchung und Forschung, 202*, 15–18.

Virtanen, A. J. D., Goedeken, L., & Tong, C. H., (1997). Microwave assisted thawing of model frozen foods using feed-back temperature control and surface cooling. *Journal of Food Science, 62*(1), 150–154.

Volden, J., Borge, G. I. A., Hansen, M., Wicklund, T., & Bengtsson, G. B., (2009). Processing (blanching, boiling, steaming) effects on the content of glucosinolates and antioxidant-related parameters in cauliflower (*Brassica oleracea* L. ssp. botrytis). *LWT-Food Sci. Technol., 42*, 63–73.

Wang, J., Xu-Hai, Y. A. S. Mujumdar, D. W., Jin-Hong, Z., Xiao-Ming, F., Qian, Z., Long, X., Zhen-Jiang, G., & Hong-Wei, X., (2017). Effects of various blanching methods on weight loss, enzymes inactivation, phytochemical contents, antioxidant capacity, ultrastructure and drying kinetics of red bell pepper (*Capsicum annuum* L.). *LWT – Food Science and Technology, 77*, 337–347.

Wang, R., Zhang, M., Mujumdar, A. S., & Jin-Cai, S., (2009). Microwave freeze-drying characteristics and sensory quality of instant vegetable soup. *Drying Technology, 27*, 962–968.

Wang, W., & Chen, G. H., (2003). Numerical investigation on dielectric material assisted microwave freeze-drying of aqueous mannitol solution. *Drying Technology, 21*(6), 995–1017.

Wang, Y., Wig, T. D., Tang, J., & Hallberg, L. M., (2003). Dielectric properties of foods relevant to RF and microwave pasteurization and sterilization.*Journal of Food Engineering, 57*, 257–268.

Wang, Y., Yuanrui, L., Shaojin, W., Li, Z., Mengxiang, G., & Juming, T., (2011).Review of dielectric drying of foods and agricultural products.*Int. J. Agric. and Biol. Eng., 4*(1), 1–19.

Wang, Y., Zhang, M., Mujumdar, A. S., & Mothibe, K. J., (2012). Microwave-assisted pulse-spouted bed freeze-drying of stem lettuce slices-effect on product quality. *Food and Bioprocess Technology, 6*(12), 3530–3543.

Wang, Z. H., & Shi, M. H., (1999). Microwave freeze drying characteristics of beef. *Drying Technology, 17*(3), 433–447.

Wappling-Raaholt, B., & Ohlsson, T., (2005). Microwave thawing and tempering. In: Schubert, H., & Regier, M., (eds.), *The Microwave Processing of Foods* (pp. 292–312). Cambridge: Woodhead.

Wojdyło, A., Figiel, A., Krzysztof, L., Paulina, N., & Jan, O., (2014). Effect of convective and vacuum-microwave drying on the bioactive compounds, color, and antioxidant capacity of sour cherries. *Food Bioprocess Technol., 7*, 829–841.

Wu, G. C., Zhang, M., & Mujumdar, A. S., (2010). Effect of calcium ion and microwave power on structural and qualitative changes in drying of apple slices. *Drying Technology, 28*(4), 517–522.

Xiao, H. W., Jun-Wen, B., Da-Wen, S., & Zhen-Jiang, G., (2014). The application of superheated steam impingement blanching (SSIB) in agricultural products processing – A review. *Journal of Food Engineering, 132*, 39–47.

Xiao, H. W., Zhongli, P., Li-Zhen, D., Hamed, M. E. M., Xu-Hai, Y., Arun, S. M., Zhen-Jiang, G., & Qian, Z., (2017). Recent developments and trends in thermal blanching – A comprehensive review. *Information Processing in Agriculture, 4*, 101–127.

Xin, Y., Min, Z., Baoguo, X. A., Benu, A., & Jincai, S., (2015). Research trends in selected blanching pretreatments and quick freezing technologies as applied in fruits and vegetables: A review. *International Journal of Refrigeration, 57*, 11–25.

Xu, Y. Y., Zhang, M., &Tu, D. Y., (2005). A two-stage convective air and vacuum freeze-drying technique for bamboo shoots.*International Journal of Food Science and Technology, 40*(6), 589–595.

Xu, Y., Yanping, C., Yaqun, C., Wenshui, X., & Qixing, J., (2016). Application of simultaneous combination of microwave and steam cooking to improve nutritional quality of cooked purple sweet potatoes and saving time. *Innovative Food Science and Emerging Technologies, 36*, 303–310.

Yang, Y., Isabel, A., & Montserrat, P., (2016).Effect of the intensity of cooking methods on the nutritional and physical properties of potato tubers.*Food Chemistry, 197*, 1301–1310.

Yongsawatdigal, J., & Gunasekaran, S., (1996). Microwave-vacuum drying of cranberries: Part 1. Energy use and efficiency. *Journal of Food Processing and Preservation, 20*, 121–143.

Zhang, D., & Hamauzu, Y., (2004). Phenolics, ascorbic acid, carotenoids and antioxidant activity of broccoli and their changes during conventional and microwave cooking. *Food Chem., 88*(4), 503–509.

Zhang, H., & Datta, A. K., (2001). Electromagnetics of microwave heating: Magnitude and uniformity of energy absorption in an oven. In: Datta, A. K., & Anatheswaran, R. C., (eds.), *Hand Book of Microwave Technology for Food Applications* (pp. 33–63).

Zhang, M., Hao, J., & Rui-Xin, L., (2010). Recent developments in microwave-assisted drying of vegetables, fruits, and aquatic products—drying kinetics and quality considerations. *Drying Technology, 28*, 1307–1316.

Zhang, M., Jiang, H., & Lim, R. X., (2010). Recent developments in microwave-assisted drying of vegetables, fruits, and aquatic products—drying kinetics and quality considerations. *Drying Technology, 28*, 1307–1316.

Zhang, M., Tang, J. M., & Mujumdar, A. S., (2006). Trends in microwave-related drying of fruits and vegetables. *Trends in Food Science and Technology, 17*(10), 524–534.

Zhang, W., Liu, F., Nindo, C., & Tang, J., (2013). Physical properties of egg whites and whole eggs relevant to microwave pasteurization. *J. Food Eng., 118*, 62–69.

Zhang, W., Luan, D., Tang, J., Sablani, S., Rasco, Lin, H., & Liu, F., (2015). Dielectric properties and other physical properties of low-acyl gellan gel as relevant to microwave assisted pasteurization process. *J. Food Eng., 149*, 195–203.

Zhang, W., Tang, J., Liu, F., Bohnet, S., & Tang, Z., (2014). Chemical marker M2 (4-hydroxy-5-methyl-3(2H)-furanone) formation in egg white gel model for heating pattern determination of microwave-assisted pasteurization processing. *J. Food Eng., 125*, 69–76.

Zhang, X., Kishore, R., Hanwu, L., Roger, R., & Brajendra, K. S. (2017). An overview of a novel concept in biomass pyrolysis: Microwave irradiation. *Sustainable Energy Fuels*. doi: 10.1039/c7se00254h.

Zhao, Y., Yajun, J., Baodong, Z., Weijing, Z., Yafeng, Z., & Yuting, T., (2017). Influence of microwave vacuum drying on glass transition temperature, gelatinization temperature, physical and chemical qualities of lotus seeds. *Food Chemistry, 228*, 167–176.

Zhong, X., Kirk, D. D., & Eva, A., (2015). Effect of steamable bag microwaving versus traditional cooking methods on nutritional preservation and physical properties of frozen vegetables: A case study on broccoli (Brassica oleracea). *Innovative Food Science and Emerging Technologies, 31*, 116–122.

Zhong, Y., Tu, Z., Liu, C., Liu, W., Xu, X., Ai, Y., & Wu, J., (2013). Effect of microwave irradiation on composition, structure and properties of rice (*Oryza sativa* L.) with different milling degrees.*Journal of Cereal Science, 58*(2), 228–233.

Zielinska, M., & Michalska, A., (2016). Microwave-assisted drying of blueberry (*Vacciniumcorymbosum* L.) fruits: Drying kinetics, polyphenols, anthocyanins, antioxidant capacity, color and texture. *Food Chemistry, 212*, 671–680.

Zielinska, M., Sadowski, P., & Błaszczak, W., (2015). Freezing/thawing and microwave-assisted drying of blueberries (*VacciniumcorymbosumL.*). *LWT – Food Science and Technology, 62*, 555–563.

# CHAPTER 3

# Effects of Drying Technology on Physiochemical and Nutritional Quality of Fruits and Vegetables

DEEPAK KUMAR VERMA,[1] MAMTA THAKUR,[2]
PREM PRAKASH SRIVASTAV,[1] VAHID MOHAMMADPOUR KARIZAKI,[3]
and HAFIZ ANSAR RASUL SULERIA[4]

[1]*Agricultural and Food Engineering Department,*
*Indian Institute of Technology Kharagpur, Kharagpur, West Bengal, India,*
*E-mails: deepak.verma@agfe.iitkgp.ernet.in; rajadkv@rediffmail.com*
*(D.K. Verma); pps@agfe.iitkgp.ernet.in (P.P. Srivastav)*

[2]*Department of Food Engineering and Technology, Sant Longowal*
*Institute of Engineering and Technology, Longowal, Punjab, India,*
*E-mails: thakurmamtafoodtech@gmail.com; mamta.ft@gmail.com*

[3]*Chemical Engineering Department, Quchan University of Advanced*
*Technology, Quchan, Iran, E-mail: mohammadpour_vahid@yahoo.com*

[4]*UQ Diamantina Institute, Translational Research Institute, Faculty*
*of Medicine, The University of Queensland, Woolloongabba, Brisbane,*
*Australia, E-mail: hafiz.suleria@uqconnect.edu.au*

## 3.1 INTRODUCTION

Since fruits and vegetables of diverse cultivars are seasonally produced, therefore, they may not remain fresh round the year. Thus, these food materials are usually processed by various drying techniques. The fresh produce, when dried or concentrated usually, has lower *MC* (Chang et al., 2016). The fruits and vegetables being perishable in nature can't preserve for several months because a great amount of nutrients and moisture is available for microbial growth. The drying of fruits and vegetables is the only technique for preserving them by inhibiting the growth of microorganisms (Lijuan et al., 2005).

Drying is one of the oldest methods of preservation simply means the removal of water from food and is currently extensively employed in food processing (Ratti, 2001). The process of drying involves the interaction between several internal and external factors like process variables, kind of raw materials, etc. (May et al., 1999). The drying process aims at reducing the water activity ($a_w$) so that the foodstuffs can be kept for longer durations at the room temperature (Bonaui et al., 1996).

There are several drying methods ranging from the old ones like sundrying, hot air (HA) drying (HAD) to the recently explored techniques such as osmotic dehydration (OD), ultrasound-assisted osmotic dehydration (UAOD), microwave-assisted HAD, etc. which are the main attraction for researchers (Toğrul and Pehlivan, 2004; Alibas, 2007; Kumar and Tiwari, 2007; Pu and Sun 2017).

Numerous factors such as type of raw material, available equipments, consumer demand for high-quality end-product, economical, and environmental conditions, etc. are considered while selecting the dryer for a particular operation (Sablani, 2006). Initially, the drying process was concentrated only on the process and technological characteristics with the goal to increase the shelf-life *(SL)* of food materials. No attention was given to maintain the quality-related attributes. However, nowadays, the scenario has changed, and the food industry has changed its focus from quantity to quality products, and many research and development projects are being carried out in the drying of food materials to produce products with high quality.

The consumers today are demanding the convenience health products that mimic the properties of fresh ones. It is a great challenge for food researchers and industry to produce such high quality dried products (Sablani, 2006; Mujumdar and Huang, 2007).

Since the drying technique involves the transient energy as well as momentum or mass transfer via a medium with phase change and rarely with chemical reactions, it is very complicated to interpret the process having complex modeling (Mujumdar and Huang, 2007). Therefore, numerous changes ranging from microstructural modification to volume deformation and from the reduction of nutrients to their degradation are associated with the drying of food materials.

Several investigations have been carried out to examine the characteristics of foodstuffs as affected by the drying process. For instance, the food quality attributes during the intermittent drying are examined by Kumar et al. (2014), and many aspects like physical modifications, color, ascorbic acid, sugar, β-carotene, and caffeine level were measured. Another research conducted by Rathnayaka et al. (2017) focused on the alterations in the morphology of food tissues during drying and also reviewed the current advances in the numerical modeling of foodstuffs. The effect of vacuum and conventional drying on the color attributes of potato, banana, apple, and carrot was reported by Krokida et al. (1998) whereas Zogzas et al. (1994) examined the influence of drying on the physical characteristics like particle density, bulk density, porosity, and shrinkage of potato and apple cubes to determine the quality of food materials.

Keeping above in view, this book chapter aims to evaluate the impact of drying on the physicochemical parameters of fruits and vegetables like texture, shrinkage, color, porosity, flavor, etc. Further, the changes in the nutritional properties during drying are also highlighted. This chapter also focused on the organoleptic and rehydration characteristics of fruits and vegetables as influenced by the drying process.

## 3.2 DRYING: AN OVERVIEW

Fruits and vegetables are important agricultural products to preserve human beings have always been of great importance. Various technologies are employed to preserve the fresh produce, out of which drying is most valuable and economical. Based on the heat and mass transfer laws, the drying is a thermo-physical action that involves the heat penetration inside the product resulting in the release of water vapors (via evaporation) (Gavrila et al., 2008). The drying of these agricultural products has the one major objective, i.e., to lower the moisture of commodity to a safe level so that deteriorative reactions and microbial spoilage can be prevented in order to enhance the *SL* of dried products (Akpinar and Bicer, 2004; Gürlek et al., 2009). The role of water activity ($a_w$) is also significant in the preservation of fruits and vegetables, which are capillary-porous food; therefore, it is important to reduce $a_w$ below 0.7 to inhibit the microbiological growth. The microbial proliferation and chemical reactions altering the food chemistry are generally reduced because of the maintenance of minimal water activity and reduced free available moisture in dried fruits and vegetables and their products thus increasing their *SL* (Khattab and Barakat, 2002; Akpinar and Bicer, 2004; Perumal, 2007; Gürlek et al., 2009; Sagar and Suresh, 2010).

During the drying process, there are basically two important mechanisms—constant rate period and falling rate period (Alonge and Adeboye, 2012). The first mechanism (also called a drying flux period) involves the evaporation of water present above the surface into the atmosphere. The relative humidity (RH) of HA typically decreases on drying; therefore, the air absorbs more moisture and removes the unbound water present at the surface. At this stage, the surface water remains constant due to the quick transfer of internal water towards the surface for balancing the moisture (Heldman and Haptel, 1999), and the product temperature almost resembles the wet-bulb temperature of air because of the phenomena of evaporative cooling (Traub, 2016). The major driving forces of drying may be determined on the basis of the difference between the surface water vapor pressure of food ($P_w$) and drying air. Several factors affecting the drying rate include the air velocity, temperature, initial MC, RH, and surface area of food to be dried, and the heat supply is the limiting factor (Heldman and Haptel, 1999; Hawlader et al., 2004). During the constant rate period, the drying rate is calculated using equation (1) (Fortes and Okos, 1980):

$$dM/dt = h_T A_P (T_a - T_{wb})/L \qquad (1)$$

where,

$dM/dt$ = Drying rate;

$h_T$ = Convective heat transfer coefficient;

$A_P$ = Product surface area;

$T_a$ = Dry-bulb temperature of dried air;

$T_{wb}$ = Wet-bulb temperature of air above the surface;

$L$ = Heat of water vaporization.

Numerous research works showed the actual drying occurring during the falling rate period (Gürlek et al., 2009). When drying proceeds, insufficient free moisture leads to failure of maintaining the maximum drying rate, and critical MC is reached. Due to insufficient movement of bound water to the product surface, the rate of surface moisture loss is greater than replenished moisture or bound water (Heldman and Haptel, 1999;

Traub, 2002), resulting in declination of the rate of drying. This point is known as falling rate period and limiting factor here is case hardening (Katekawa and Silva, 2007a, b). The parameters influencing the drying rate are collectively known as diffusion coefficient in equation (2) given by Mujumdar (1991):

$$M - M_e / M_0 - M_e = e^{-kt} \qquad (2)$$

where,

$M$ = Moisture content;

$M_e$ = Equilibrium moisture content (EMC);

$M_0$ = Initial moisture content;

$k$ = Drying constant;

$t$ = Time.

The drying process involves the water removal from food materials resulting in some changes in the physicochemical and nutritional composition like color, flavor, nutrients, etc. (Chua et al., 2000a, b, 2002; Mayor and Sereno, 2004; Sablani, 2006; Sagar and Suresh, 2010). Since vitamins like A, C, and B$_1$ are heat sensitive and are prone to oxidative degradation; huge loss of such vitamins has been observed during drying. According to Musa et al. (2010), the drying process reduces the protein, β-carotene and vitamin C in tomatoes (Musa et al., 2010). Joshi and Mehta (2010) reported 430/100 g of leaf powder decrease in anti-nutritional factor (oxalate) in drumstick (*Moringa oleifera*) after drying. Drying has attained great importance in food industry to reduce the post-harvest losses in fruits and vegetables as well as to improve the product quality that can be accomplished using the suitable drying technique such as mechanical drying, freeze-drying (FD), vacuum drying, thermal drying and chemical drying (Hawlader et al., 2005; Sobukola et al., 2007; Gavrila et al., 2008; Bala and Janjai, 2009).

### 3.2.1 DIFFERENT TYPE OF DRYING

#### 3.2.1.1 CONVECTIVE DRYING (CD)

The drying techniques like fluidized bed and spouted bed drying and tray drying are included under the convective drying (CD), which exhibited the analogous anthocyanin levels, color, taste, and rehydration of halved cranberries (Grabowski et al., 2002). As compared to tray drying, the fluidized bed drying including the spouted beds has the potential to reduce drying time and increase the efficiency, if vibration is incorporated to it (Grabowski et al., 2002). According to Grabowski et al. (2002), the berry was perfectly dried rather than approximately 50% drying (reported in earlier studies) at 90°C. Blueberries was dried 5 times faster at 70C using the spouted bed dryer (*Vaccinium corymbosum*) compared to tray trying (Devece et al., 1999). Another investigation by Feng et al. (1999) also reported the increased drying rate and end-product quality in spouted bed dryer during the pre-drying and drying of blueberries at 70°C where no damage took place to fruits when compared to the freeze and vacuum drying (Feng et al., 1999).

Onions are commercially dried using the CD technology where constant drying is less effective than three drying stages to ensure the improved sensory properties of onion flakes. However, there are some negative effects of convectional drying on the product quality which includes the reduced rehydration properties and excess thermal damage (Holdsworth, 1986). The four-stage drying is more effective and can reduce the drying time of onion by 25% compared to the two-stage drying process (Munde et al., 1988). Since CD is a time-consuming process, numerous undesired changes may occur in the end product Lewicki, (1998a, b). For example, a 3 mm thick onion slice when dried at 76°C under their velocity of 27 m/min, 58 minutes are required, as reported by Akbari et al. (2001, 2006).

The browning rate was increased by 2–3 (% db) MC while air velocity didn't influence the product browning and thiosulfinate content significantly (Kaymak-Ertekin and Gedik, 2005). The drying temperature intensively affects the sugars, acidity, and vitamin C content of fruits and vegetables without significantly influencing the fat, ash, crude protein, and fiber (Mota et al., 2010). The fluidized bed drying may prevent the scorching or the agglomeration of onion pieces than tray, tunnel, or conveyer belt drying (Gelder, 1962; Gummery, 1977). The drying temperature below 53°C preserved the green color of chopped spring onion, revealed by Swasdisevi et al. (1999).

### 3.2.1.2   FREEZE DRYING (FD)

FD of blueberries retained the color, vitamin A, B3, and C-contents, low bulk density, and high rate of rehydration as compared to other drying techniques such as convective air, micro-convection, and vacuum oven drying (Yang and Atallah, 1985). The excellent taste and color properties, rehydration capacity, and anthocyanin content were reported in cranberries by FD by Grabowski et al. (2002). Similarly, the FD of onion resulted in minimum degradation of color, flavor, and nutrient content with retention of rehydration attributes than traditional drying methods (Hamed and Foda, 1966; Popov et al., 1976) whereas the hot drying of onions followed by freezing revealed the similar rehydrating attribute, reduced volume and deeper color of product which makes them more suitable for storage, packaging, and transportation (Andreotti et al., 1981). The FD is the best technique for preserving the quality and sensory properties of food; however, it has a limitation of high cost (Somogyi and Luh, 1986; Freeman and Whenham, 2006).

### 3.2.1.3   MECHANICAL DRYING

Mechanical drying is a technique to remove moisture forcefully either by ambient air or through heated air by the following two mechanisms, they are: A) *Heated air drying*: It deals with the elevated temperatures resulting in rapid drying, which continues until the desired MC is obtained. Different types of dryers like batch, continuous flow, and re-circulating type batch dryer are widely used for this purpose. B) *Low-temperature drying:* It offers better control to maintain the temperature and suitable moisture content (*MC*) than the sun drying. It can be used at any time of day and employ less labor, especially in the case of re-circulating dryers.

### 3.2.1.4   MICROWAVE DRYING (MWD)

There is a reduction of drying time and product quality in the microwave drying (MWD), as reported by Beaudry et al. (2003). Cranberries, dried under microwave, resulted in poor sensory attributes due to burn berries, blackened surface color and unpalatable flavor (Beaudry et al., 2003). The onions when freeze and microwave dried had similar characteristics except the improved rehydration in FD (Kamoi et al., 1981). Recently, Abbasi and Azari, (2009) revealed the microwave-vacuum-freeze dryer as a simple, effective, economic, and innovative method to quickly dry the fruit and vegetables which has the commercial significance.

### 3.2.1.5   SUN DRYING

The fruits and vegetables are the rich source of nutrients in area which have favorable climatic conditions, however, they are found in limited amount in semi dry-arid areas where sun drying is most important to ensure the make availability of fruits and vegetables in dry season also (Maeda and Salunkhe, 1981; Mosha et al., 1997; Mulokozi and Svanberg, 2003; Sagar and Suresh, 2010). Sun drying is a traditional and most ancient method (Figure 3.1) and works as an economic energy source having no instrumental cost (Salunkhe, 1974; Kinabo et al., 2004). The drawback of sun drying is its poor quality of drying and the susceptibility of food to get contaminated by dust insects and birds (Salunkhe, 1974; Al-Juamily, 2007; Bala and Janjai, 2009; Sagar and Suresh, 2010) and unhygienic due to present of microorganisms and insects such as flies (Kinabo et al., 2004). The main drawback is loss of nutrients due to direct exposure of sunlight, especially ultra-violet radiation (UVR). Trade potential and worth of the crop get affected due to loss of food quality during drying (Gürlek et al., 2009).

**FIGURE 3.1**   Sun drying of green vegetables with direct exposure of sunlight. (*Source*: Reprinted with permission from Verma et al., 2017. © Apple Academic Press.).

### 3.2.1.6   VACUUM DRYING

Vacuum drying is an excellent technique among all drying techniques, and provides low energy and capital cost for the final product (Yang and Atallah,

1985). It led to the better retention of quality properties of blueberry when compared to FD (Yang and Atallah, 1985). According to the study by Grabowski et al. (2002), vacuum dried cranberries had better rehydration properties, anthocyanin content, color, and taste than the direct heating techniques. Numerous investigations recommended the combination of freeze and vacuum drying more economic for industrial processing than other prevalent drying methods (Yang and Atallah, 1985; Grabowski et al., 2002; Azoubel and Murr, 2003). Mitra et al. (2011) reported the significant increase of moisture diffusivity along with the increase of temperature and thickness which varied from 1.32 E−10 to 1.09 E−01 $m^2$/s and 32 E−10 to 1.09 E−09 $m^2$/s for treated and untreated onion samples, respectively.

### 3.2.2 WHY DRYING IS IMPORTANT IN FRUITS AND VEGETABLES

Drying of fruits and vegetables after harvesting is the most effective technique to reduce the post-harvest losses (Wiriya et al., 2009). The losses after harvest are enormous (30–40%) resulting into nutritional and economic losses (Kimambo, 2007). The Indian food processing sector has still not fully grown, therefore nearly 40% fruits and vegetables are lost every year and only less than 10% are processed (MAFC, 2009). Several studies reported the solar drying as a simple inexpensive technique which has great potential to minimize the post-harvest losses and ensure the round-year availability of fresh produce (Habou et al., 2003; Mujumdar, 2004). But fruits and vegetables when exposed to solar radiations resulted in the degradation of vit. A and C under the atmospheric oxygen and unde-sired sensory properties (Kabasa et al., 2004; Barret, 2007). Moreover, the performance of solar dryer under different climatic conditions may affect the retention of nutrients, sensory attributes, local standards, and SL of the end products. Due to the diverse weather, the solar-dried fruits and vegetables varied in the final quality in Tanzania (Ringo, 2008). Therefore, the weather conditions also decide the market opportunities and product development of solar-dried fruits and vegetables.

As discussed above, there is a strong need to develop the appropriate technology for the processing of fruits and vegetables in order to retain the nutrients which improve health. The efforts must be taken by the food industry and economy to reduce the post-harvest losses quantitatively and qualitatively (Kabasa et al., 2004). The technology which would preserve the nutrients, improve quality, safety, and *SL* of fresh and processed fruits and vegetables is essential to gain the market access (Temu et al., 2008) and the outcomes of such technology should be served as a guide to establish the quality standards, optimum processing, and storage conditions. It can also provide the base to develop the economic technology-dependent solutions of fruits and vegetable processing to enhance food security. In the long term, it may serve as a guide to establish a food processing enterprise for fruits and vegetables drying, thus generating the employment for youths and women.

## 3.3　EFFECTS OF DRYING ON DIFFERENT PHYSIOCHEMICAL AND NUTRITIONAL QUALITY PARAMETERS

Drying technology—one of the oldest methods of fruits and vegetables preservation (Sobukola et al., 2007), which is based on the removal of water to minimize the chemical and microbial deterioration and extend the *SL* of dried products (Perumal, 2007). It also reduces the weight and volume; so that the storage, packaging, and transportation costs can be minimized (Sagar and Suresh, 2010). After harvesting the produce like fruits and vegetables, sun drying is the most common method in developing countries, which is however, related to several problems (Folaranmi, 2008). During sun drying, the product can be degraded by the following factors:

1. *Biotic factors* including the infestation by insects, rodents, and other animals; and
2. *Abiotic factors* such as rain, storm, windborne dirt, dust, etc.

Thus, the end product of sun drying is of poor quality which doesn't meet the quality and safety guidelines of national and international authorities (Ivanova and Andonov, 2001), whereas the mechanical drying such as cyclone drying, drum drying, and FD produce the quality products but they are costly and hazardous to environment. Among all techniques, the solar energy is highly significant maintaining the quality and safety of products which must be encouraged at industrial level too (Eltief et al., 2007).

The term quality in food refers to the ability to satisfy the stated needs of consumers based on the features and characteristics of a product (CFSAN, 2006). The quality of fruits and vegetables during the drying is greatly affected due to the increased temperature (heat) and exposure time resulting to the physicochemical and biochemical changes (Jimoh et al., 2008; Jokić et al., 2009). Simply, the deterioration of quality can be divided into *physical quality*, *chemical quality*, and *nutritional quality,* as described in Figure 3.2 (Methakhup, 2003; Methakhup et al., 2005; Perera, 2005). Such quality changes also rely on any pre-treatment given to fruits and vegetables prior the drying process to minimize the negative impacts of quality during drying and storage of fruits and vegetables. These pre-treatments encourages the cell destruction or damage to enzymatic routes which later on cease the metabolism of cut fruits and vegetables tissues (Lewicki, 2006).

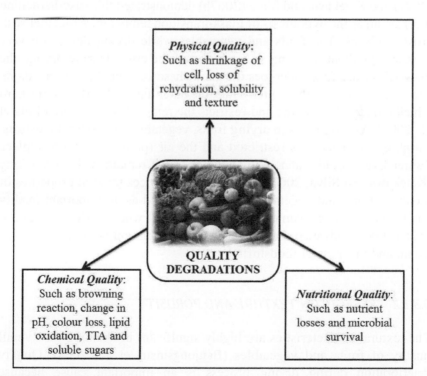

**FIGURE 3.2**    Quality degradations during drying and storage of fruits and vegetables.

### 3.3.1  PHYSICAL QUALITY

#### 3.3.1.1  EFFECTS ON SHRINKAGE

The shrinkage is simply the deformations in volume of materials like fruits, vegetables, and other foodstuffs during the drying process (Katekawa and Silva, 2007a). The drying rate of fruits and vegetables is mainly affected by the diffusion coefficient that is based on drying and diffusion coefficient is influenced by shrinkage (Lima et al., 2002). The volume deformation or reduction is nearly equal to the removed liquid water volume if material is undergone the ideal shrinkage (Waje et al., 2005). Several investigations reported the strong correlation between the drying conditions of fruits and vegetables and their shrinkage extent which are referred as case-hardening phenomenon (Ratti, 1994; Mayor and Sereno, 2004). Later on, Rahman (2001) and Katekawa and Silva (2007b) demonstrated this case-hardening mechanism as the hypothesis of glass-transition temperature together with crust synthesis. The fruits and vegetables while drying develop a very high MC gradient resulting in formation of the crust. During drying, the material surface first undergoes the glass phase keeping the material damp and rubbery towards the center. Thus, the outer material surface turns hard which reduces the volume and restricts the removal of moisture (Talla et al., 2004). As progresses in drying fruits, vegetables, and other foodstuffs, shrinking of material is restricted and the air (pores) is used to replace the removed liquid water; due to which, crust remain without cracking (Katekawa and Silva, 2007b). Shrinkage influences textural properties of the dried fruits and vegetables as it is attributed as an important quality parameter for the consumers. The experimental work carried by Funebo et al. (2000) confirmed the mechanism of association between shrinkage extent and firmness of foodstuffs.

#### 3.3.1.2  EFFECTS ON TEXTURE AND POROSITY

The textural characteristics are highly significant in deciding the overall quality of fruits and vegetables (Banjongsinsiri et al., 2004). Quality deterioration during drying process is an important cause because of the textural properties changes in fruits and vegetable like solid foods (Fellows, 2009). Textural properties which include chewiness, crispness, crunchiness, fibrousness, firmness, tenderness, etc. parameters are measured by applying the force to the produce (UNCTAD, 2007). Fruits and vegetables undergo collapse of structure during drying process allow to lead the firmer textures and increased chewiness (Gabas et al., 2007). During drying, there is the accumulation of internal stress due to the starch gelatinization, cellulose crystallization, pectin degradation, and localized differences in MC which lead to the injury and destruction of rigid cells (Fellows, 2009). This texture loss makes the food material to look like shrunken shrivel and is responsible for the variation in the texture quality of dried food (Maskan, 2001).

Several studies reported the poor texture of fruits and vegetables due to the higher drying rate and temperature which caused the heat damage or injury to the plant tissues (Quintero-Ramos et al., 1992). The fruit tissues are mildly affected on drying at a moderate rate and low temperature. Two other factors viz. structure of cellular tissues and their composition also play a great role in the texture of fruits and vegetables (Bhale, 2004).

The texture is also directly correlated to porosity (Aguilcra and Stanlcy, 1999), which refers to the fractions of voids or pores in a material and also provides us information regarding the degree of shrinkage (Rahman, 1995). The degree of shrinkage also measures the shape and size which determine the quality of dried products including color (Ayrosa and Nogueira, 2003; Perera, 2005).

### 3.3.2 CHEMICAL QUALITY

#### 3.3.2.1 EFFECTS ON COLOR

Color—the most relevant attribute of quality of dried foods is directly associated with the consumer's acceptability (Methakhup, 2003; Methakhup et al., 2005; Bonazzi and Dumoulin, 2011). During drying, many chemical and biochemical reactions may change the color of fruits and vegetables and this problem has always been associated with drying (Maskan et al., 2002). Their long-term storage may lead to discoloration due to the browning reactions (Methakhup, 2003; Methakhup et al., 2005). There are generally two types of browning reaction in food, i.e., enzymatic and non-enzymatic.

Enzymatic browning is a significant biochemical reaction in fruits and vegetables which have adverse effects on several quality properties such as color, flavor, taste, and nutrients. In enzymatic browning, an enzyme "polyphenol oxidase (PPO)" catalyzes the oxidation reaction of polyphenols and convert them into "melanin"—the brown pigment (Guerra et al., 2010). The phenolic compounds (PCs) are oxidized and then polymerized to form "melanin" during the drying and storage of many fruits and vegetables like apples, bananas, and potatoes (Perera, 2005). This type of browning in these fruits and vegetables can also be seen upon tissue bruising, cutting, peeling, diseased, and exposed to any number of abnormal conditions. When exposed to air, the quick darkening of injured fruits and vegetables tissues occurs because of the synthesis of brown pigments "melanins" from the PCs (Manzocco et al., 2001). Fruits like citrus, bananas, melons, mushrooms, olives, peaches, plums, roots, pears, tea, etc. contain the enzyme "phenolase" that includes several other enzymes such as ascorbinase, catecholase, phenoloxidase, and tyrosinase (Wiriya et al., 2009). According to reports, more than 50% of estimated losses in fruits and vegetables occur due to enzymatic browning only (Whitaker and Lee, 1995); therefore, there is a great need to understand and control the diphenolase enzymes in fruits and vegetables. The knowledge of enzymatic browning and their control is essential to increase the global trade of fruits and vegetables (Marshall et al., 2000). Many chemical treatments and blanching may prevent the enzymatic browning reactions by inactivating the enzymatic activity (Gupta et al., 2002; Hossain and Bala, 2002). The PPO enzyme can be inactivated at temperatures above 60°C. According to Wakayama, (1995), the PPO enzyme activity decreased from 49–13% to an undetectable level in Fuji apple on increasing the temperature from 50–60°C to more than 70°C. Roig et al. (1999) also reported the non-enzymatic browning during the storage of citrus juice in which the ascorbic acid, its isomers, and derivatives were employed to inhibit the enzymatic browning.

There are three following types of non-enzymatic browning (Klieber, 2000; Perera and Baldwin, 2001; Methakhup, 2003; Methakhup et al., 2005):

1. Maillard reaction;
2. Caramelization; and
3. Ascorbic acid degradation.

The chemical reaction between the amine group (-NH$_2$) of amino acids and reducing sugars at elevated temperature is known as Maillard reaction (Figure 3.3); which is the basis for the production of off-color and flavor compounds thus deteriorating the product quality (Miao and Roos, 2006). The fruits undergo the condensation seen in Maillard reaction and oxidation of ascorbic acid and their derivatives lead to the browning (Barreiro et al., 1997; Maskan, 2001). The thermal decomposition of ascorbic acid occurring with or without the aerobic conditions and amino-compounds is termed as ascorbic acid browning (Wedzicha, 1984). The non-enzymatic reaction rate is influenced by several factors like MC, time, and temperature of drying, pH, reactants characteristics and concentration of fruit and vegetable (Manzocco et al., 2001; Belitz et al., 2004). For example, the rate of Maillard reaction is increased with the temperature increase and interestingly, also with higher sugar content (Chua et al., 2002).

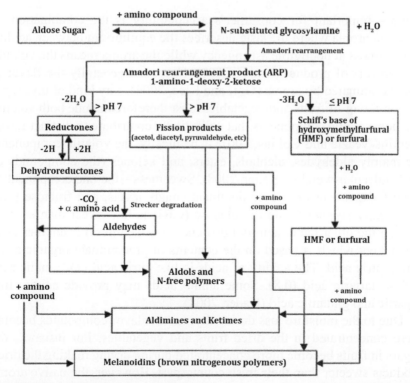

**FIGURE 3.3**  Maillard browning pathway.
(*Source*: Reprinted from Hodge, 1953).

There are several reactions in addition to the ones described above which results into the browning and affect the color of fruits and vegetables during drying. Among them, the pigment degradation especially the degradation of carotenoids and chlorophyll is the most common together with ascorbic acid and PCs oxidation (Barreiro et al., 1997; Maskan, 2001; Manzocco et al., 2001; Maskan et al., 2002). The chemical changes in carotenoids and chlorophyll are mainly induced by the heat and oxidation during drying process. The pigment loss can be high when fruits and vegetables are dried at higher temperatures for longer times (Ersus and Yurdagel, 2007; Araújo et al., 2017). Other factors influencing the color of fruits and vegetables include the pH, acidity, variety and heavy metal contamination (Maskan, 2001).

## 3.3.2.2 EFFECTS ON FLAVOR

Flavor—important for consumer's acceptability is the combination of taste and aroma where the taste describes the equilibrium between acidity and sourness at negligible astringency while the aroma means the volatile components of product (Kader, 2008). However, generally the flavor is found dominated by aroma (Goff and Klee, 2006). Any kind of investigation on flavor of fruits and vegetables must therefore include both volatile and non-volatile components that collectively contribute to distinct flavor (Perkins-Veazie and Collins, 2001). Chemically, the volatile components are mainly aldehydes, alcohols, esters, and ketones (the compounds of low-molecular-weight) (Kader, 2008). Sweetness—the one component of taste is dependent on the predominant sugars levels, and fructose (1.2) is typically sweeter while the glucose (0.64) is less sweet than sucrose (1). On the other hand, acidity (also known as sourness) is another taste component which is based on the contents of predominant organic acids w.r.t., citric acid. The acidity of acids is like citric acid (1) > malic acid (0.9) > tartaric acid (0.8). Some amino acids may provide acidity like aspartic and glutamic acid (Kader, 2008).

Due to the moisture loss during drying, the flavor compounds become more concentrated in the dried fruits and vegetables. For instance, the sugars in fruits become more concentrated during drying making the dried products sweeter than their fresh counterpart. However, the native aroma during drying is considered to reduce by auto-oxidation of unsaturated fatty acids or heat-based Maillard reaction leading to the formation of additional volatile flavor compounds (VFCs) (Goff and Klee, 2006). Such VFCs have boiling points lower than the water; and interestingly, VFCs having relatively higher diffusivity and volatility are escaped during the initial stages of drying (Steinke et al., 1989; Feng et al., 1999; Yaylayan and Roberts, 2001; Poll and Thi, 2004). Few volatile compounds are lost at later stages (Kaewtathip and Charoenrein1, 2012; Raice et al., 2015). During storage, the major factors resulting in the aroma loss are oxidation of lipids, pigments, and vitamins, and deterioration rate is affected by temperature and water activity ($a_w$) of storage. It is, therefore, essential to regulate the drying conditions properly to reduce the drying losses.

### 3.3.2.3 EFFECTS ON PHYTOCHEMICALS

Phytochemicals simply refer to the bioactive non-nutrient components of plants like fruits and vegetables, cereals, grains, and other plant-derived foods that are found in different plant parts in varying forms and concentrations (Liu, 2003). The plant when exposed to radiations, extremely high temperatures and some injury may increase the concentration of phytochemicals (Shahidi and Naczk, 2004). Chemically, the phytochemicals are synthesized by the amino acids like tyrosine and phenylalanine and they act as anti-oxidant and protectant against the ultraviolet (UV) light (Shahidi and Naczk, 2004). Among several phytochemicals the total phenolic compounds (TPCs) are the most significant.

### 3.3.2.3.1 Total Phenolic Compounds (TPCs)

The amount of polyphenolic compounds in fruits and vegetables are totally dependent on the cultivars, agronomic conditions like type of soils, fertilizers, temperature, and cultivation methods, post-harvest storage, transportation, and further secondary processing techniques (Bennett et al., 2011). According to a study by Hamroun-Sellami et al. (2013), the TPCs are significantly affected by the different drying methods and their levels may be raised or lowered with minor or no modifications. However, another study by Chang et al. (2006) and Suvarnakuta et al. (2011) reported the sharp reduction of TPCs levels due to the phenols degradation on drying the vegetable leaves using the oven, microwave, and sun.

The oxidative reactions may also degrade the PCs in fresh produce due to PPO enzyme resulting in inter-molecular condensation reactions thus reducing their level Bennett et al. (2011). Likewise, several researchers have reported the decrease in the TPCs when apricots, persimmons, mulberry, and ginger leaves, shiitake mushroom, olive mill waste, sweet potatoes, tomatoes, and prune were dried (Dewanto et al., 2002; Caro et al., 2004; Chang et al., 2006; Choi et al., 2006; Park et al., 2006; Madrau et al., 2008; Obied et al., 2008; Chan et al., 2009; Katsube et al., 2009; Mao et al., 2010). Several other researchers have shown the significant impact of drying on the TPCs amount of fruits and vegetables (Zhang et al., 2012; Hamrouni-Sellami, 2013). The findings of Segura-Carretero et al. (2010) suggested that the composition of PCs varies with diversity of fruits and vegetables cultivars among the other factors like maturity stage and exposure to sunlight. The similar variations in TPCs composition and concentration are also exhibited by the dried fruits such as apricot, mango, and palm (Uhlig and Walker, 1996; Piga et al., 2005; Ribeiro et al., 2007; Boetang et al., 2008; Madrau et al., 2008; Vega-Galves et al., 2009).

### 3.3.2.3.2  Phytate Content

Phytate or inositol hexakisphosphate (Insp6) is a major constituents of plant tissues, where is considered as a phosphate source (Phillippy et al., 2004). Phosphorous as phytate is found in a considerable amount in different plant organs. Phytate is the complex salt of magnesium or calcium with myo-inositol and it has potential to precipitate the minerals based on which the phytates may reduce the bioavailability of iron, calcium, magnesium, and zinc (Ravindran et al., 1994; Phillippy et al., 2004).

The phytates are found to reduce the triglycerides, cholesterol, and blood clots thus reducing the risk of heart disease. They are also helpful in inhibiting the development of renal stone and also removing the traces heavy metal ions from human body (Kumar et al., 2010).

Ravindran et al. (1994) examined the phytate content of different fruits and vegetables of tropical origin. Also, the average content between 0.5% and 1% phytate is reported for raw seeds, while many starchy foodstuffs including potatoes, cassava, and yam contain about 0.5–0.10% phytate on wet basis (w.b) (Phillippy et al., 2004). The phytate level in plant tissue is found to reduce during some food unit operations such as milling and soaking (Mahgoub and Elhag, 1998). Also, certain thermal or biological treatments such as appertisation cause phytate level to be decreased (Lestienne et al., 2005). The phytate content of four sorghum cultivars were analyzed during the different treatments (e.g., milling, soaking, and fermentation) applied in food production.

Generally, the studies determining the effect of drying on the phytate level is very scanty, however, the phytate concentration is found reduced in drying.

### 3.3.2.3.3   Pigments

A pigment refers to the natural coloring compound that is used as a natural colorant to while preparing the foods like jams, jellies, confectionery, preserves, frozen desserts, etc., nutraceuticals, functional foods, and pharmaceutical products (Ersus and Yurdagel, 2007). The process of drying modifies the color in foodstuffs and may result in the degradation of pigments.

Tang and Chen (2000) prepared the freeze-dried carotenoid powder and studied the stability of pigment during storage. During carrot juice processing, the major by-product is carrot pulp waste that is rich in α-and β-carotene which were analyzed by authors with photodiode-array detection by HPLC.

Oberoi and Sogi (2015) investigated the pigments in spray- and freeze-dried watermelon juice powder from three cultivars and found a higher *a* and *b* value and lower L value in freeze-dried juice powder than the spray-dried one. However, a good correlation between the colorimetric values and lycopene content was exhibited by the spray-dried powder.

Lee et al. (2015) studied the conventional drying of tea leaves and evaluated the responses of lyophilic pigments like chlorophyll a and b, β-carotene, pheophytin, and lutein which have numerous positive effects on health. The pigment values were also compared with other drying processes like freeze-, oven-, and MWD, which revealed that the concentration of pigment metabolites are affected by the drying method. De Souza et al. (2015) examined the color stability of spray-dried pigment derived from by-products of red grape including seeds and peels which are rich source of anthocyanins. The plant tissues contain a large amount of anthocyanins whose retention during drying is affected by the concentration of the carrier agent and temperature (Ersus and Yurdagel, 2007; De Souza et al., 2015).

For the evaluation of color, numerous color order systems are developed including the Munsell, Lovibond, hunter lab, CIE, and CIELAB system. Commonly, the CIE *lab* (CLELAB) color measurement system is used to interpret the colors during drying. In this system, the location of any color is characterized by color coordinates including *l*, *a* and *b* where lightness and darkness is shown by *l* value, *a* value represents the difference between red and green while the difference between blue and yellow is exhibited by *b* value. Further, the total color change ($\Delta E$), which is visual by human eyes also can be derived using the following equation (3) (Sahin and Sumnu, 2006; Karizaki et al., 2013):

$$E^* = \sqrt{(l^* - l_0^*)^2 + (a^* - a_0^*)^2 + (b^* - b_0^*)^2} \qquad (3)$$

where, the color values of raw or control samples are referred to as $l_0$, $a_0$, and $b_0$ components.

### 3.3.3  NUTRITIONAL QUALITY

Drying, like all methods of preservation, can result in the loss of some nutrients (Kendall et al., 2004). However, studies have shown that solar drying retained the maximum nutrients in fruits and vegetables than other drying methods. During drying as well as storage of dried products, the time-temperature combinations generally decide the huge variations in their nutritional value (Morris et al., 2004). While preparing the fruits and vegetables, the losses are extremely high, which sometimes may exceed those caused by drying (Morris et al., 2004; Karim et al., 2008). During washing, peeling, and blanching of fresh produce, the major proportion of water-soluble nutrients are lost (Bruhn et al., 2007). As per the research conducted by Mason et al. (2001), some of the salts, sugars, and water-soluble vitamins are typically lost while preparation and blanching of vegetables significantly reduces the vitamin C (47.1% loss), Ca, Na, Mg, K, P, Fe, and Zn (Mepba et al., 2007). The effects of drying on the macro- and micro-nutrients are discussed in detail under the following sub-section. Macronutrients are essential nutrients required in relatively large amounts, and they include carbohydrates, fats, and proteins. They provide energy and chemical building-blocks for tissues (McKinley Health Center, 2008).

### 3.3.3.1   EFFECTS ON CARBOHYDRATE CONTENT

The carbohydrates are the major component found in plant foods accounting for more than 90% of their dry matter (FAO, 1995). The total carbohydrate level varies from a little to2% in cucumber and squashes to more than30% in sweet potatoes (Khan, 1990). In fruits, the carbohydrate content ranges from 1.5–2.6%, and interestingly, no starch is present in the ripe fruits because it is converted to the simple sugars during the ripening process. The fruits mainly contain sucrose, fructose, and glucose as simple sugars. The fruits also contain dietary fiber like vegetables (Lintas, 1992; Lester et al., 2006; Cordain, 2012). On the other hand, the vegetables contain simple sugars, starch, and dietary fiber as major carbohydrates, and their amount varies from 27% (sweet potato) to 1–2% (leafy and stem vegetables). Further, the dietary fiber also ranges from 0.8–8.0% in cucumber and artichoke, respectively (Lintas, 1992).

During the drying of fruits and vegetables, the main chemical transformation includes the changes in carbohydrates level (Perera, 2005). There is an increase in the macronutrients like carbohydrates, proteins, and fats. The heat treatment may induce the Maillard browning reaction resulting in the flavor modification due to the interaction of amino acids with reducing sugars. According to a study by Hassan et al. (2007), the carbohydrate contents were significantly lowered in dried fruits and vegetables than the fresh samples. However, some studies exhibited no change in the fiber and energy content of fruits and vegetables because fiber is not heat-labile (Kendall et al., 2004; Perera, 2005; Barret, 2007).

### 3.3.3.2   EFFECTS ON FAT CONTENT

The fruits and vegetables generally contain the minute amount of lipids which may later on lead to rancidity and objectionable odors due to the formation of ketones, acids, and hydro-peroxides during the oxidation of unsaturated fatty acids (Perera, 2005). Therefore, rancidity is the major concern in foods during drying carried out at higher temperature which causes greater oxidation of fats compared to lower temperatures. In such cases, the antioxidants are thought to be highly effective to prevent the oxidation of fats.

### 3.3.3.3   EFFECTS ON FIBER CONTENT

There are two main categories of dietary fibers may be found in foodstuffs. The first group includes soluble fiber that is found in various varieties of fruits and vegetables. Gum and pectin are the examples of soluble fiber. The insoluble fiber comprises of second group which includes the cellulose, hemicellulose, and lignin, found mainly in fruits having edible seeds and beans (Langrish, 2008). In addition to these fibers, some foods also contain the partly-soluble fiber fractions (Kutoš et al., 2003).

The soluble fibers are beneficial in the growth of natural bacteria in the digestive system thus boosting our immune system whereas the insoluble fibers are also beneficial in the improvement of colon and gut health (Langrish, 2008).

According to recent investigations, the dietary fibers protect the human against cardiovascular diseases, colon cancer, diabetes, obesity, and other diverticular diseases (Lee et al., 1992; Kutoš et al., 2003).

In addition, the dietary fibers have several interesting physical characteristics, such as viscosity increasement, water, and oil holding, gel formation, and swelling (Borchani et al., 2011). Each fiber component has distinct physiological effect; therefore, the detailed knowledge of fiber content during drying of fruits and vegetables is essential because, processing may alter the fiber content, mainly its physico-chemical composition. For instance, the beans when dried reduced the total dietary fiber (TDF) including soluble and insoluble dietary fiber (Kutos et al., 2003). According to Borchani et al. (2011), date fibers when freeze-dried had maximum values of water holding, oil holding and swelling capacities compared to oven drying and sun drying. Langrish (2008) extracted the cellulosic fibers from citrus fruits and studies their drying behavior.

### 3.3.3.4   EFFECTS ON PROTEIN CONTENT

The fruits and vegetables sadly, don't contain the good quality and quantity protein because of the amino acid imbalance and lack the essential amino acid in required amounts. Therefore, they have very low value of protein efficiency ratio (Spiller, 2001). The fruits and vegetables are generally considered as "incomplete proteins" or sources of "low biological value protein." The fruits based diets are therefore too low in protein (Mahan and Escott-Stump, 2000). Most fruits contain generally less than 1% protein but some vegetables may contain much higher protein content than fruits. Approximately 5% of per capita proteins are contributed by leguminous vegetables and this protein is generally high in quality due to the presence of required amount of essential amino acids (Khetarpal and Kochar, 2011).

### 3.3.3.5 EFFECTS ON VITAMIN CONTENT

Vitamins are necessary to obtain from diet due to the inability of human body to synthesize them (Yeung and Laquatra, 2003). They play a great role in carrying the physiological and biological activities, metabolism, and maintenance of body (Anderson and Young, 2008). The vitamins are required in minute amounts for their catalytic action in addition to growth, maintenance, and other important functions (FAO/WHO, 2003). The vitamins based on their solubility can be classified into two main groups: (a) *Fat soluble vitamins* (Vit. A, D, E, and K) and (b) *Water soluble vitamins* (Vit. B and C). The fat-soluble vitamins in excess amount are stored in liver therefore not required daily for nutrition whereas the water soluble vitamins are eliminated through urine from body making their consistent supply necessary (Berdanier and Zempleni, 2009). The vitamins are found in good amount in both fresh and dried fruits and vegetables however, some vitamins are heat labile and lost during drying (Araújo et al., 2017). However, it is essential to understand the drying effect on such important nutrients.

### 3.3.3.5.1 Ascorbic Acid (Vitamin-C)

As discussed above, the ascorbic acid (Figure 3.4) is water-soluble, can't be stored in the body and therefore, its regular supply is necessary to maintain the health. It is beneficial in the synthesis of cartilages, tendons, skin, and blood vessels, wound healing, and preventing the scurvy, a condition where collagen synthesis is damaged (Richard and Ferrier, 2011). Ascorbic acid acts as an anti-oxidant, immune-system modulator and improves the iron absorption in body (Shrimpton, 1993; Cook and Reddy, 2001).

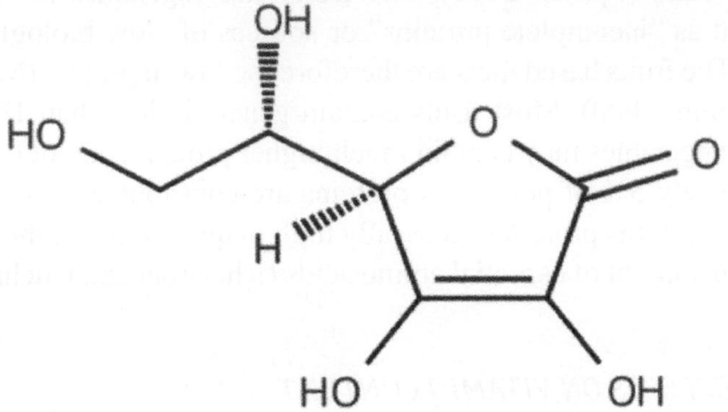

**FIGURE 3.4**   Structure of L-ascorbic acids.

The fruits and vegetables like broccoli, cabbage, capsicum, cauliflower, kiwifruits, honeydew melon, mango, oranges, papaya, peas, potatoes, strawberries, spinach, tomatoes, etc. typically supply nearly 90% of ascorbic acid (in the human diet) (Lee and Kader, 2000), but, its amount is affected by variations in cultivars, agro-climatic conditions, soil, maturity stage and harvesting methods (Lee and Kader, 2000; Podsedek, 2007). The plant tissue, exposed to increased light intensity during growth may contain the higher levels of vitamin C (Lee and Kader, 2000). This vitamin is the most heat sensitive poorly stable and easily destroyed during handling, processing, and storage. The ascorbic acid is extremely vulnerable to light, oxidation, water, pH, heat, dissolved oxygen, and metallic ions such as $Ag^+$, $Cu^{2+}$, and $Fe^{3+}$ (Lin et al., 1998; Lee and Kader, 2000). The destruction of vitamin C may lead to the quality loss and color formation therefore; the drying of fruits and vegetables affected the color and ascorbic acid content.

Depending on environmental conditions, the ascorbic acid degradation (Figure 3.5) includes two main steps, i.e., anaerobic and aerobic degradation. The first mechanism of degradation is very complex and of no value in food products. While the aerobic degradation mechanism oxidizes the ascorbic acids to dehydroascorbic acid (DAA) under an aerobic condition. The further hydrolysis and oxidation of DAA form the diketogulonic acid ($C_6H_8O_7$) and furfural. Here, the DAA has vitamin C activity and also undergoes an irreversible reaction under anaerobic and aerobic conditions on heating, similar to ascorbic acid (Ottaway, 1993). Further polymerization of furfural form the brown color pigments which affect the color, flavor, and nutritional properties of fruits and vegetables during drying.

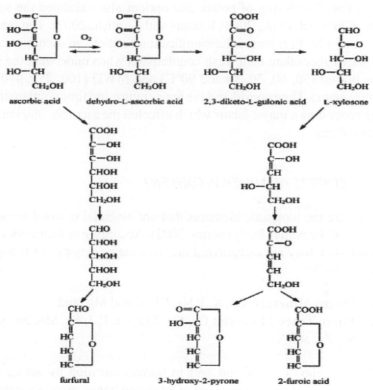

**FIGURE 3.5**   Degradation of ascorbic acid.
(*Source*: Reprint from Yuan and Chen, (1998)).

After the process of drying, the retained amount of ascorbic acid determines the end quality of dried produce (Marfil et al., 2008; Lee and Kader, 2000). The fruits on sun-drying lost 84% of vitamin C, but when covered with polyethylene, the solar-dried fruits reduced only 71% ascorbic acid (Ndawula et al., 2004). Its degradation is more during slow drying as compared to quick drying. Further, maximum ascorbic acid is retained during the freeze and spray drying of fruit or vegetable tissues compared to HA and solar drying (Jayaraman and Das Gupta, 1995). Leong and Oey (2012) found that ascorbic acid content is changed by 125, 90 and 99% in pepper (which in fresh state contain 10.8 mg/g ascorbic acid) after drying for 10 min at 98°C, FD for 48 hours and frozen-thawing at 4°C for 2 hours, respectively. According to the studies of Özkan et al. (2007), the drying time has strong association with the degradation of ascorbic acid from leaves. The drying of potato and apricot also exhibited the similar results (Khraisheh et al., 2004; Karatas and Kamişli, 2007). The findings of Ali et al. (2016) reported the significant alterations in ascorbic acid of dried guava slices than their fresh counterparts when underwent the solar, freeze, hot-air (50, 60, 70, 80 and 90°C) and MWD (100, 250, 440, 600 and 1000 watts). They considered the temperature and time combination of drying process as a major factor which affected the ascorbic acid contents in guava slices.

### 3.3.3.6   EFFECTS ON MINERAL CONTENT

Minerals are the inorganic elements that are essential to usual metabolic processes of the human body (Sonni, 2002). According to their importance in human diet, they are categorized into two major classes as (Yeung and Laquatra, 2003):

1.  **Macro Minerals:** Ca, K, P Na, Cl, S, and Mg; and
2.  **Micro/Trace Minerals:** Cu, Co, Mb, Cr, F, I, Fe, Mn, Ni, Se, Si, Sn, Zn, and V.

Minerals play a significant part in human nutrition by acting as an indispensable structural component of teeth and bones, enzymes cofactors and catalysts in numerous biological processes (Sonni, 2002). They may contribute to the contraction and conduction of nerve impulses (Anhwange et al., 2005). Some heavy metals like arsenic (As), lead (Pb) and cadmium (Cd) are harmful to health and may cause acute to chronic biochemical disorders when consumed (Duruibe et al., 2007).

The fruits and vegetables are the rich sources of minerals essential for health (Barrett, 2007). Mineral content is not much affected by the drying (Kresic, 2004), while the organic and mineral constituents amount was found to increase during drying (Aliero and Abdullahi, 2009). Similarly, the study by Hassan et al. (2007) showed the enhancement of mineral elements except sodium while drying. The retention of minerals are better in solar drying compared to sun drying in tomatoes and *Vernonia amyg-dalina* (bitter leaf) (Aliero and Abdullahi, 2009; Babarinde et al., 2009).

### 3.3.4 SENSORY QUALITY

The handling of fresh fruits and vegetables after harvesting is a great challenge due to their perishable nature and drying process may increase the storage life and prevent the microbial growth of these products, as discussed above (Forero et al., 2015). However, the process of drying has negative impact on the sensory properties of dried products and in some cases, the fruits and vegetables may either lose or damage or degrade the sensory characteristics completely due to the chemical reactions like oxidation and enzymatic interactions during the process (Forero et al., 2015; Da Costa Ribeiro et al., 2016).

The evaluation of sensory properties during the process is greatly significant which can be achieved by design a panel test to examine the consumer products. Several investigations recommended the use of panel test to measure the sensory characteristics during drying (Van Ruth and Roozen, 1994; Choi et al., 2017). For instance, the van Ruth and Roozen (1994) trained a panel of critically selected 24 sensory analysts varying from age 20 to 65 years to analyze the capsicum after drying. The sensory evaluation of onion powders produced by CD was performed by 12 panelists (7 female, 5 male, aged 24–30) (Choi et al., 2017). In another study by Mokhtarian et al. (2017), ten trained and experienced panelists evaluated the sensory attributes of pistachio during shade, solar, and sun drying.

The empirical methods together with statistical techniques are employed for sensory evaluation of food materials where the basic human senses like smell, sight, and touch are used for analysis. Numerous parameters may be considered for sensory evaluation based on the food product. In the case of drying of pistachio, the shell appearance, shell splitting, and finger releasing kernel were the most important characteristics, reported in the study by Mokhtarian et al. (2017).

Choi et al. (2017) considered the odor attributes of onion as sensory properties of dried onion powders while the conventional HAD. The major aroma features of onions are due to the presence of different aldehydes which contain six to nine carbons ($C_6$–$C_9$), and quickly escaped during drying.

Reviewing the existing literature on drying of foodstuffs, the following sensory characteristics were considered as the most relevant: aroma, taste, appearance, color, texture, flavor, and mouthfeel (Coutron-Gambotti et al., 1999; Carvalho et al., 2017). Additionally, the difference among various products can be determined by overall palatability and acceptability like dried pistachio when dried using different methods varied greatly in their overall palatability where fruits which are undergone traditional sun drying and air recycling enjoyed the highest sensory score than other treatments (Mokhtarian et al., 2017).

Each drying method has distinct impact on the sensory characteristics of food products. For example, there will be loss of aroma compounds and chemical degradation during HAD while the OD, freeze, and spray drying may produce products having acceptable sensory quality (Forero et al., 2015; Jiang et al., 2017). Moreover, the recent combined drying techniques like UAOD, and microwave-vacuum drying technologies make the product having an improved sensory score (Karizaki et al., 2013; Da Costa Ribeiro et al., 2016; Cano-Lamadrid et al., 2017; Jiang et al., 2017). For instance, the overall sensory acceptance score of pear that is osmotic dehydrated using HA method was obtained 17% higher than the only conventionally dried product (Da Costa Ribeiro et al., 2016).

### 3.3.5 REHYDRATION QUALITY

For utilizing the dried products, the rehydration is an important unit operation which includes the moisturization of dried products so as to regain the characteristics similar to their fresh counterparts. This process can be performed from room temperature to boiling conditions for 2 min–24 hours where water mass: dry material mass ranges from 1:5 to 1:50. Rehydration water is still or maybe occasionally or continuously stirred (Lewicki, 1998a, b). The most common index applied to express rehydration of dry food tissue is the following ratio:

$$\frac{weight\ after\ rehydration}{initial\ weight}$$

Various researchers have named this ratio to rehydration capacity, rehydration ratio (RR), hydration coefficient, and percent water absorption (Lewicki, 1998a, b).

Numerous studies have suggested the different modes for performing the rehydration process (Vergeldt et al., 2014; Ricce et al., 2016). Simply, the dried product can be immersed in distilled water at ambient temperature and evolution of food moisture with time is calculated using the mass balance. After sometime, the dried fruits or vegetables are removed from water, drained, dried superficially using a paper towel, weighed, and then finally returned to the water.

While the drying process, the rehydration behavior can be studied as an induced damage like integrity loss and reduced rehydration ability of the foodstuff. When the structure of food tissue remains integrated, then the dried food material, theoretically, must absorb and gain moisture equal to the water content of initial foodstuff before drying (Marques et al., 2009).

In this case, the fruit and vegetable tissues may undergo the compositional and structural changes due to the drying process, process conditions, and different pre-treatments before drying including blanching, OD, and sonication which influence the quality as well as rehydration behavior of food. According to Diaz et al. (2003), there was significant destruction of sensory properties during traditional air drying (AD) at elevated temperatures or for longer times, which thus reduces the rehydration potential of dried products.

Recently, the researchers are paying great attention to the rehydration process of fruits and vegetables; for instance, the rehydration capacities were studied by Maskan, (2001). He reported the decreased rehydration capacity of microwave dried kiwifruit slices as compared to the HA, and air-microwave (AM) dried slices. Further, the water absorption rate was also faster in microwave dried samples compared to other drying techniques. Rehydration behaviors of different freeze-dried fruits (e.g., mango, guava, pineapple, papaya, and acerola) were considered by Marques et al. (2009). They related rehydration process to glass transition temperature, and the structural changes during the process. The authors also found the proper mathematical models for description of the rehydration kinetics. Rehydration quality of carrot was considered by Lin et al. (1998), showing no significant difference in texture and overall acceptability for carrot samples which had undergone the HA drying, vacuum-MWD, and FD and rehydrated for 10 min at 100°C. A summary of studies directly dealt with the dehydration process is listed in Table 3.1.

**TABLE 3.1**   A Summary of Reports on Rehydration Process of Dried Fruits and Vegetables

| Fruits and Vegetable | Experimental Conditions | | | | Main Result(s) | References |
|---|---|---|---|---|---|---|
| | Drying Process | | Rehydration Process | | | |
| | Dryer and Types of Drying | Drying Parameters | Rehydration Time (min) | Rehydration Temperature | | |
| Tomato | Rotary tray dryer | T: 45–60°C V = 0.6, 1.2 m/s | 50 | Room temperature | The rehydration ratio values reported in this study varied from 3.7±0.4 to 4.8±0.6. | Santos-Sánchez et al., 2012 |
| Apple | Convective drying (CD) | T: 70°C V = 1.3 m/s | - | Room temperature | The rehydration process plasticizes the glassy matrix of dried apples. | Lewicki et al., 1997 |
| Sour cherry | Conventional drying, hybrid drying (HD) | T: 50–70°C M = 120–180 W | 120 | 30°C | The higher rehydration ability was found from hybrid (microwave-conventional) drying. | Horuz et al., 2017 |
| Mushroom | Convective drying, microwave-vacuum drying | T: 50–70°C M = 115–285 W P = 6.5–23.5 KPa | 0–180 | 30, 100°C | Rehydration behavior of mushrooms was significantly affected by the pressure of system. | Giri and Prasad, 2007 |
| Date palm | Air drying | T: 50–80°C V = 1.5 m/s | - | 15–45°C | Rehydration of dried date palm samples recorded higher activation energy, and lower diffusivity than air-drying. | Falade and Abbo, 2007 |
| Orange | Combined microwave-air drying | T: 60°C V = 2 m/s M = 0–0.88 W/g | - | 25°C | Rehydration behavior of orange samples was modeled through Weibull's and Peleg's equations. | Díaz et al., 2003 |

**TABLE 3.1**  *(Continued)*

| Fruits and Vegetable | Experimental Conditions | | | | Main Result(s) | References |
|---|---|---|---|---|---|---|
| | Drying Process | | Rehydration Process | | | |
| | Dryer and Types of Drying | Drying Parameters | Rehydration Time (min) | Rehydration Temperature | | |
| Spinach | Microwave drying | M = 180–900 W | 300 | 30–70°C | The activation energy for rehydration process was determined using Arrhenius equation and obtained as 23.84 kJ/mol. | Dadali et al., 2008 |
| Hawthorn | Convective drying | T: 50–70°C V = 0.5–1.3 m/s | 540 | 20–80°C | The rehydration ratio was increased with increasing air velocity and air temperature. | Aral and Beşe, 2016 |
| Strawberry | Freeze drying | T: –30–25°C | 5–120 | Room temperature | A decrease in rehydration capacity was found in the case of freeze-dried strawberry samples that were osmotically dehydrated in osmotic solution. | Agnieszka and Andrzej, 2010 |
| Pumpkin | Air drying, combined method | T: 50–70°C M = 105–315 W | 0–200 | - | The pumpkin pieces dried by conventional air drying shows a higher rehydration than pieces dried by combined method. | Seremet et al., 2016 |

**TABLE 3.1** *(Continued)*

| Fruits and Vegetable | Experimental Conditions | | | | Main Result(s) | References |
|---|---|---|---|---|---|---|
| | Drying Process | | Rehydration Process | | | |
| | Dryer and Types of Drying | Drying Parameters | Rehydration Time (min) | Rehydration Temperature | | |
| Blueberry | Microwave-assisted drying | T: 60, 80°C | - | 21°C | In comparison with single-stage hot air drying and single-stage microwave vacuum drying, hot air pre-drying step with microwave-assisted drying led to higher rehydration capacity. | Zielinska and Markowski, 2016 |
| Carrot | Freeze drying | T: –30–25°C | - | Room temperature | The rehydration kinetics of freeze-dried carrot samples was measured by magnetic resonance imaging and time-domain NMR as a function of thermal pretreatment. | Vergeldt, van Dalen et al., 2014 |
| Carrot | Freeze drying | T: –30–25°C | - | Room temperature | The effects of freeze-drying operation on rehydration properties of carrot were investigated. | Voda, Homan et al., 2012 |

T: Temperature;
V: Air velocity;
M: Microwave power;
P: Pressure.

Furthermore, several researchers have utilized and described the mechanisms related to the enhancement of the rehydration process of fruits and vegetables (Ricce et al., 2016). For example, Ricce et al. (2016) showed that the ultrasound waves can enhance the rehydration rate of carrot slices at low temperatures. The authors performed the drying process at 40 and 60°C on the pieces were pretreated with ultrasonic waves *(UW)*. It was concluded that the pre-treated samples had higher rehydration rate in comparison with the untreated carrot slices. Also, the Peleg model was applied to describe the rehydration kinetics of the carrot.

Generally, the aim of using new or hybrid drying (HD) techniques such as ultrasound-assisted, microwave, microwave-vacuum, and microwave-air drying is to obtain faster or/and better rehydration.

## 3.4   CONCLUSION AND SUMMARY

Shrinkage, texture, and porosity are the most common and important physical characteristics of a foodstuff while the color and flavor are the most common chemical characteristics. In this study, the physicochemical properties of fruits and vegetables were investigated as influenced by the process of drying. As well, the nutritional quality of these kinds of food materials during drying process was studied. For this, the effects of drying operation on carbohydrate, fat, fiber, protein, vitamin, and mineral content of fruits and vegetables were reviewed. Furthermore, the sensory and rehydration characteristics of dried fruits and vegetables were considered affecting their overall acceptability.

It was shown that the drying methods may partially or totally affect the quality of a product. In the other words, different changes in chemical, physical, and/or nutritional characteristics of food materials may occur during the drying process. As a result, the different aspects of a food product such as color, shape, structure, texture, composition, etc., may be changed, decreased, or lost. Additionally in the recent years, there have been several studies on the application of new or HD techniques such as microwave, microwave-vacuum, microwave-air drying for obtaining faster and better rehydration, and higher sensory quality in comparison with the conventional methods. In the one hand, using different drying techniques, dried fruits and vegetables can be found to show completely different attributes. On the other hand, there is a need to produce dried fruits and vegetables with better sensory characteristics. Thus, it is expected to develop combined drying technologies or create new drying methods in the near future.

## KEYWORDS

- air drying
- amine group
- food preservation
- food quality
- hybrid drying
- Millard reaction
- nutritional property
- physical property
- shelf-life
- total phenolic compounds

## REFERENCES

Abbasi, S., & Azari, S., (2009). Novel microwave-freeze drying of onion slices. *International Journal Food Science and Technology, 44*, 974–979.

Agnieszka, C., & Andrzej, L., (2010). Rehydration and sorption properties of osmotically pretreated freeze-dried strawberries. *Journal of Food Engineering, 97*(2), 267–274.

Aguilera, J. M., & Stanley, D. W., (1999). *Microstructural Principles of Food Processing and Engineering* (2nd edn., pp. 325–372). Aspen Publishers, Gaithersburg, USA.

Akbari, S. H., & Patel, N. C., (2006). Optimization of parameters for good quality dehydrated onion flakes. *Journal Food Science and Technology, 43*, 603–606.

Akbari, S. H., Patel, N. C., & Joshi, D. C., (2001). Studies on dehydration of onion. *ASAE International Annual Meeting* (p. 130). Held at Sacramento, California, USA.

Akpinar, E. K., & Bicer, Y., (2004). Modeling of the drying of eggplants in thin-layers. *International Journal of Food Science and Technology, 39*(1), 1–9.

Ali, M. A., Yusof, Y. A., Chin, N. L., & Ibrahim, M. N, (2016). Effect of different drying treatments on color quality and ascorbic acid concentration of guava fruit. *International Food Research Journal, 23*, S155–S161.

Alibas, I., (2007). Microwave, air and combined microwave–air-drying parameters of pumpkin slices. *LWT—Food Science and Technology, 40*(8), 1445–1451.

Aliero, A. A., & Abdullahi, L., (2009). Effect of drying on the nutrient composition of *Vernonia amygdalina* leaves. *Journal of Phytology*, *1*, 28–32.

Al-Juamily, K. E. J., Khalifa, A. J. N., & Yassen, T. A., (2007). Testing of the performance of a fruit and vegetable solar drying system in Iraq. *Desalination, 209*, 163–170.

Alonge, A. F., & Adeboye, O. A., (2012). Drying rates of some fruits and vegetables with passive solar dryers. *International Journal of Agricultural and Biological Engineering*, *5*(4), 83–90.

Anderson, J., & Young, L., (2008). *Water-Soluble Vitamins*. URL: http://www.ext.colostate.edu/pubs/foodnut/09312.html (Accessed on 8 November 2019).

Andreotti, R., Tomasicchio, M., & Macchiavelli, L., (1981). Freeze drying of carrots and onions after partial air drying. *Ind. Conserve, 54*(2), 87–91.

Anhwange, B. A., Ajibola, V. O., & Oniye, S. J., (2005). Composition of bulk, trace and some rare earth elements in the seeds of *Moringa olefera* (Lam), *Deutarium microcapium* (Guill and Perr) and *Bauhinia moandra* (Kurz). *Journal of Food Technology, 3*(3), 290–293.

Aral, S., & Beşe, A. V., (2016). Convective drying of hawthorn fruit (*Crataegus* spp.): Effect of experimental parameters on drying kinetics, color, shrinkage, and rehydration capacity. *Food Chemistry, 210*, 577–584.

Araújo, A. C., Oliveira, S. M., Ramos, I. N., Brandão, T. R. S., Monteiro, M. J., & Silva, C. L. M., (2017). Evaluation of drying and storage conditions on nutritional and sensory properties of dried Galega kale (*Brassica oleracea* L. var. Acephala). *Journal of Food Quality Volume*, 1–9.

Ayrosa, A. I. B., & Nogueira, P. R., (2003). Influence of plate temperature and mode of rehydration on textural parameters of precooked freeze-dried beef. *Journal of Food Processing and Preservation, 27*(3), 173–180.

Azoubel, P. M., & Murr, F. E. X., (2003). Effect of pre-treatment on the drying kinetics of cherry tomato (*Lycopersicon esculentum* var. cerasiforme). In: Welti-Chanes, J., Velez-Ruiz, F., & Barbosa-Cánovas, G. V., (eds.), *Transport Phenomena in Food Processing* (pp. 137–151). CRC Press, New York, USA.

Babarinde, G. O., Akande, E. A., & Anifowose, F., (2009). Effects of different drying methods on physico-chemical and microbial properties of tomato (*Lycopersicon esculentum* Mill) var. Roma. *Fresh Produce, 3*(1), 37–39.

Bala, B. K., & Janjai, S., (2009). Solar drying of fruits, vegetables, spices, medicinal plants and fish: Developments and Potentials. *International Solar Food Processing Conference* (p. 180). India.

Baranowski, J. D., & Nagel, C. W., (1982). Inhibition of pseudomonas flurisecens by hydroxycinnamic acids and their alkyl esters. *Journal of Food Science, 47*(5), 1587–1589.

Barreiro, J. A., Milano, M., & Sandoval, A. J., (1997). Kinetics of color change of double concentrated tomato paste during thermal treatment. *Journal of Food Engineering, 33*, 359–371.

Barrett, D. M., (2007). Maximizing the nutritional value of fruits and vegetables. *Journal of Food Technology, 61*(4), 40–44.

Beaudry, C., Raghavan, G. S. V., & Rennie, T. J., (2003). Microwave finish drying of osmotically dehydrated cranberries, *Drying Technology, 21*(9), 1797–1810.

Belitz, H. D., Grosch, W., & Schieberle, P., (2004). *Food Chemistry* (3rd reversed and extended edition, p. 1070). Springer, Berlin.

Bennett, L. E., Jegasothy, H., Konczak, I., Frank, D., Sudharmarajan, S., & Clingeleffer, P. R., (2011). Total polyphenolics and anti-oxidant properties of selected dried fruits and relationships to drying conditions. *Journal of Functional Foods, 3*, 115–124.

Berdanier, C., & Zempleni, J., (2009). *Advanced Nutrition: Macronutrients Micronutrients and Metabolism* (p. 163). CRC Press, New York, USA.

Bhale, S. D., (2004). Effects of ohmic heating on color, rehydration and textural characteristics of fresh carrot cubes. *M.Sc. Dissertation, Department of Biological and Agricultural Engineering*. Louisiana State University, Louisiana, USA.

Boetang, J., Verghese, M., Walker, L. T., & Ogutu, S., (2008). Effect of processing on antioxidant contents in selected dry beans (*Phaseolus vulgaris* L.). *LWT—Food Science and Technology, 41*, 1541–1547.

Bonaui, C., Dumoulin, E., Raoult-Wack, A. L., Berk, Z., Bimbenet, J. J., Courtois, F., Trystram, G., & Vasseur, J., (1996). Food drying and dewatering. *Drying Technology, 14*(9), 2135–2170.

Bonazzi, C., & Dumoulin, E., (2011). Quality changes in food materials as influenced by drying processes. In: Tsotsas, E., Arun, S., & Mujumdar, A. S., (eds.), *Drying Technology: Product Quality and Formulation* (1st edn., Vol. 3, pp. 1–20). Wiley-VCH Verlag GmbH & Co. KGaA. Published.

Borchani, C., Besbes, S., Masmoudi, M., Blecker, C., Paquot, M., & Attia, H., (2011). Effect of drying methods on physico-chemical and antioxidant properties of date fiber concentrates. *Food Chemistry, 125*(4), 1194–1201.

Bruhn, C. M., Rickman, J. C., & Barrett, M. D., (2007). Nutritional comparison of fresh, frozen and canned fruits and vegetables. Part Vitamins C and B and phenolic compounds. *Journal of the Science of Food and Agriculture, 3*(1), 243–246.

Cano-Lamadrid, M., Lech, K., Michalska, A., Wasilewska, M., Figiel, A., Wojdyło, A., & Carbonell-Barrachina, Á. A., (2017). Influence of osmotic dehydration pre-treatment and combined drying method on physico-chemical and sensory properties of pomegranate arils, cultivar Mollar de Elche. *Food Chemistry, 232*, 306–315.

Caro, A. D., Piga, A. P., Pinna, I., Fenu, P. M., & Agabbio, M., (2004). Effect of drying conditions and storage period on polyphenolic content, antioxidant capacity, and ascorbic acid of prunes. *Journal of Agriculture and Food Chemistry, 52*, 4780–4784.

Carvalho, M. J., Perez-Palacios, T., & Ruiz-Carrascal, J., (2017). Physico-chemical and sensory characteristics of freeze-dried and air-dehydrated yogurt foam. *LWT—Food Science and Technology, 80*, 328–334.

CFSAN (Center for Food Safety and Applied Nutrition), (2006). *Registration for Food Facilities*. URL: http://www.cfsan.fda.gov/~furls/ovffreg.html (Accessed on 8 November 2019).

Chan, E. W. C., Lim, Y. Y., Wong, S. K., Lim, K. K., Tan, S. P., Lianto, F. S., & Yong, M. Y., (2009). Effects of different drying methods on the antioxidant properties of leaves and tea of ginger species. *Food Chemistry, 113*(1), 166–172.

Chang, C. H., Lin, H. Y., Chang, C. Y., & Liu, Y. C., (2006). Comparisons on the antioxidant properties of fresh, freeze-dried and hot air-dried tomatoes. *Journal of Food Engineering, 77*, 478–485.

Chang, S. K., Alasalvar, C., & Shahidi, F., (2016). Review of dried fruits: Phytochemicals, antioxidant efficacies, and health benefits. *Journal of Functional Foods, 21*, 113–132.

Choi, S. M., Lee, D. J., Kim, J. Y., & Lim, S. T., (2017). Volatile composition and sensory characteristics of onion powders prepared by convective drying. *Food Chemistry, 231,* 386–392.

Choi, Y., Lee, S. M., Chun, J., Lee, H. B., & Lee, J., (2006). Influence of heat treatment on the antioxidant activities and polyphenolic compounds of shitake (*Lentinus edodes*) mushroom. *Food Chemistry, 99,* 381–387.

Chua, K. J., Hawlader, M. N. A., Chou, S. K., & Ho, J. C., (2002). On the study of time varying temperature drying – effect on drying kinetics and product quality. *Drying Technology, 20*(8), 1559–1577.

Chua, K. J., Ho, J. C., Mujumdar, A. S., Hawlader, M. N. A., & Chou, S. K., (2000a). Convective drying of agricultural products—effect of continuous and stepwise change in drying air temperature. Paper No. 29. In: Kerkhof, P. J. A. M., Coumans, W. J., & Mooiweer, G. D., (eds.), *Proceedings of the 12th International Drying Symposium* (p. 230). Amsterdam: Elsevier Science.

Chua, K. J., Mujumdar, A. S., Chou, S. K., Hawlader, M. N. A., & Ho, J. C., (2000b). Convective drying of banana, guava and potato pieces: Effect of cyclical variations of air temperature on drying kinetics and color change. *Drying Technology, 18,* 907–936.

Cook, J. D., & Reddy, M. B., (2001). Effect of ascorbic acid intake on non-heme iron absorption from a complete diet. *American Journal of Clinical Nutrition, 73,* 93–98.

Cordain, L., (2012). *Sugar Content of Fruit, Fruits and Sugars.* The Paleo Diet. URL: http://thepaleodiet.com/getting-started-with-the-paleo-diet/ (Accessed on 8 November 2019).

Coutron-Gambotti, C., Gandemer, G., Rousset, S., Maestrini, O., & Casabianca, F., (1999). Reducing salt content of dry-cured ham: effect on lipid composition and sensory attributes. *Food Chemistry, 64*(1), 13–19.

Da Costa, R. A. S., Aguiar-Oliveira, E., & Maldonado, R. R., (2016). Optimization of osmotic dehydration of pear followed by conventional drying and their sensory quality. *LWT—Food Science and Technology, 72,* 407–415.

Dadali, G., Demirhan, E., & Özbek, B., (2008). Effect of drying conditions on rehydration kinetics of microwave dried spinach. *Food and Bioproducts Processing, 86*(4), 235–241.

De Souza, V. B., Thomazini, M., Balieiro, J. C. C., & Fávaro-Trindade, C. S., (2015). Effect of spray drying on the physicochemical properties and color stability of the powdered pigment obtained from vinification byproducts of the Bordo grape (*Vitis labrusca*). *Food and Bioproducts Processing, 93,* 39–50.

Devece, C., Rodríguez-López, J. N., Fenoll, L. G., Tudela, J., Catalá, J. M., Reyes, E., & García-Cánovas, F., (1999). Enzyme inactivation analysis for industrial blanching applications: Comparison of microwave, conventional, and combination heat treatments on mushroom polyphenoloxidase activity. *Journal of Agricultural and Food Chemistry, 47*(11), 4506–4511.

Dewanto, V., Wu, X., Adom, K. K., & Liu, R. H., (2002). Thermal processing enhances the nutritional value of tomatoes by increasing total antioxidant activity. *Journal of Agricultural and Food Chemistry, 50,* 3010–3014.

Díaz, G. R. Z., Martínez-Monzó, J., Fito, P., & Chiralt, A., (2003). Modeling of dehydration-rehydration of orange slices in combined microwave/air drying. *Innovative Food Science and Emerging Technologies, 4*(2), 203–209.

Duruibe, J. O. C., & Egwurugwu, J. N., (2007). Heavy metal pollution and human biotic effects. *International Journal of Physical Sciences, 2*(5), 112–118.

Eltief, S. A., Salah, A., Ruslan, M. H., & Yatim, B., (2007). Drying chamber performance of V-groove forced convective solar dryer. *Desalination, 209*, 151–155.

Ersus, S., & Yurdagel, U., (2007). Microencapsulation of anthocyanin pigments of black carrot (*Daucus carota* L.) by spray drier. *Journal of Food Engineering, 80*(3), 805–812.

Falade, K. O., & Abbo, E. S., (2007). Air-drying and rehydration characteristics of date palm (*Phoenix dactylifera* L.) fruits. *Journal of Food Engineering, 79*(2), 724–730.

FAO, (1995). *Fruit and Vegetable Processing* (p. 247). Food and Agriculture Organization of the United Nations, Rome. Italy. URL: http://www.vouranis.com/uploads/6/2/8/5/6285823/fao_fruit veg_ processing.pdf (Accessed on 8 November 2019).

FAO/WHO, (2003). *Diet, Nutrition and the Prevention of Chronic Diseases* (p. 164). World Health Organization Technical Report Series, No. 916. Geneva, Switzerland.

Fellows, P. J., (2009). *Food Processing Technology, Principles and Practice* (3ʰ edn., p. 913). Woodhead Publishing Limited. Oxford, England.

Feng, H., Juming, T., Mattinson, D. S., & Fellman, J. K., (1999). Microwave and spouted bed drying of frozen blueberries: The effect of drying and pretreatment methods on physical properties and retention of flavor volatiles. *J. Food Proc and Preserv., 23*(6), 463–479.

Feng, H., Tang, J., Mattinson, D. S., & Fellman, J. K., (1999). Microwave and spouted bed drying of frozen blueberries: The effect of drying and pretreatment methods on physical properties and retention of flavor volatiles. *Journal of Food Processing and Preservation, 23*, 463–479.

Folaranmi, J., (2008). Design, construction and testing of simple solar maize dryer. *Electronic Journal of Practices and Technologies, 13*, 122–130.

Forero, D. P., Orrego, C. E., Peterson, D. G., & Osorio, C., (2015). Chemical and sensory comparison of fresh and dried lulo (*Solanum quitoense* Lam.) fruit aroma. *Food Chemistry, 169*, 85–91.

Fortes, M., & Okos, M. R., (1980). Drying theories: Their bases and limitations as applied to foods and grains. In: Mujumdar, A. S., (ed.), *Advances in Drying* (Vol. 1, pp. 119–154). Hemisphere, Washington DC, USA.

Freeman, G. G., & Whenham, R. J., (2006). Changes in onion (*Allium cepa* L.) flavor components resulting from some post-harvest processes. *Journal of the Science of Food and Agriculture, 25*(5), 499–515.

Gabas, A. L., Telis, V. R. N., Sobral, P. J. A., & Telis-Romero, J., (2007). Effect of maltodextrin and Arabic gum in water vapor sorption thermodynamic properties of vacuum dried pineapple pulp powder. *Journal of Food Engineering, 82*, 246–252.

Gavrila, C., Ghiaus, A. G., & Gruia, I., (2008). Heat and mass transfer in convective drying processes. *Excerpt from the Proceedings of the COMSOL Conference* (p. 110). Hannover, German.

Gelder, A. V., (1962). *Fluidized Bed Process for Dehydration of Onion, Garlic and the Like* (p. 15). US Patent 3063848.

Giri, S. K., & Prasad, S., (2007). Drying kinetics and rehydration characteristics of microwave-vacuum and convective hot-air dried mushrooms. *Journal of Food Engineering, 78*(2), 512–521.

Grabowski, S., Marcotte, M., Poirier, M., & Kudra, T., (2002). Drying characteristics of osmotically pretreated cranberries: Energy and quality aspects. *Drying Technology: An International Journal, 20*(10), 1989–2004.

Gummery, C. S., (1977). A review of common onion products. *Food Trade Review, 47*(8), 452–454.

Gürlek, G., Özbalta, N., & Güngör, A., (2009). Solar tunnel drying characteristics and mathematical modeling of tomato. *Journal of Thermal Science and Technology, 29*(1), 15–23.

Hamed, M. G. E., & Foda, Y. H., (1966). Freeze drying of onions. *Z Lebensm Unters Forsh., 130,* 220.

Hamrouni-Sellami, I. M., Rahali, F., Rebey, B. I., Bourgou, S., Limam, F., & Marzouk, F., (2013). Total phenolics, flavonoids, and antioxidant activity of sage (Salvia officinalis l.) plants as affected by different drying methods. *Food and Bioprocess Technology, 6*(3), 806–817.

Hassan, S. W., Umar, R. A., Matazu, I. K., Maishanu, H. M., Abbas, A. Y., & Sani, A. A., (2007). The effect of drying method on the nutrients and non-nutrients composition of leaves of *Leptadenia hastata* (Asclipiadaceae). *Asian Journal of Biochemistry, 2,* 188–192.

Hawlader, M. N. A., Pera, C. O., & Tian, M., (2005). Influence of different drying methods on fruits' quality. In: *8ᵗʰ Annual IEA Heat Pump Conference* (p. 150). Las Vegas, Nevada, United States.

Hawlader, M. N. A., Pera, C. O., & Tian, M., (2004). Heat pump drying under inert atmosphere. In: *Proceedings of the 14ᵗʰ International Drying Symposium (IDS 2004)* (Vol. A, pp. 309–316). Sao Paulo, Brazil.

Heldman, D. R., & Haptel, R. W., (1999). *Principles of Food Processing* (p. 288). Aspen Publisher Inc, New York, USA.

Hodge, J. E., (1953). Chemistry of browning reactions in model systems. *J. Agric. Food Chem., 1,* 928–943.

Holdsworth, S. D., (1986). Advances in the dehydration of fruits and vegetables. In: McCarthy, D., (ed.), *Concentration and Drying of Foods* (pp. 293–303). London: Elsevier.

Horuz, E., Bozkurt, H., Karataş, H., & Maskan, M., (2017). Effects of hybrid (microwave-convectional) and convectional drying on drying kinetics, total phenolics, antioxidant capacity, vitamin C, color and rehydration capacity of sour cherries. *Food Chemistry, 230,* 295–305.

Hossain, M. A., & Bala, B. K., (2002). Thin layer drying characteristics for green chilly. *Drying Technology, 20*(2), 489–505.

Ivanova, D., & Andonov, K., (2001). Analytical and experimental study of combined fruit and vegetable dryer. *Journal of Energy Conversion Management, 42,* 975–983.

Jayaraman, K. S., & Das, G. D. K., (1995). Drying of fruits and vegetables. In: Mujumdar, A. S., (ed.), *Handbook of Industrial Drying* (pp. 643–689). Marcel Dekker, New York.

Jiang, N., Liu, C., Li, D., Zhang, Z., Liu, C., Wang, D., Niu, L., & Zhang, M., (2017). Evaluation of freeze drying combined with microwave vacuum drying for functional okra snacks: Antioxidant properties, sensory quality, and energy consumption. *LWT—Food Science and Technology, 82,* 216–226.

Jimoh, A. K., Olufemi, A. S. A., Biliaminu, S. A., & Okesina, A. B., (2008). Effect of food processing on glycemic response of white yam (*Dioscorea rotunda*) meals. *Original Research Article. Diabetologia Croatica, 37*(3), 111–118.

Jokić, S., Velić, D., Bilić, M., Lukinac, J., Planinić, M., & Bucić-Kojić, A., (2009). Influence of process parameters and pre-treatments on quality and drying kinetics of apple samples Czechoslovakia. *Journal of Food Science, 27*, 88–94.

Joshi, P., & Mehta, D., (2010). Effect of dehydration on the nutritive value of drumstick leaves. *Journal of Metabolomics and Systems Biology, 1*(1), 5–9.

Kabasa, J. D., Ndawula, J., & Byaruhanga, Y. B., (2004). Alterations in fruit and vegetable β-carotene and vitamin C content caused by open-sun drying, visqueen-covered and polyethylene-covered, solar-dryers. *African Health Sciences, 4*(2), 125–130.

Kader, A. A., (2008). Perspective flavor quality of fruits and vegetables. *Journal of the Science of Food and Agriculture, 88*, 1863–1868.

Kaewtathip, T., & Charoenrein, S., (2012). Changes in volatile aroma compounds of pineapple (*Ananas comosus*) during freezing and thawing. *International Journal of Food Science and Technology, 47*, 985–990.

Kamoi, I., Kikuchi, S., Matsumato, S., & Obara, T., (1981). Studies on dehydration of welsh onion and carrot by microwave. *Journal of Agriculture Science (Tokyo), 20*[th] *Anniversary Edition*, 150–168.

Karatas, F., & Kamişli, F., (2007). Variations of vitamins (A, C and E) and MDA in apricots dried in IR and microwave. *Journal of Food Engineering, 78*, 662–668.

Karim, O. R., Awonorin, S. O., & Sanni, L. O., (2008). Effect of pretreatments on quality attributes of air-dehydrated pineapple slices. *Journal of Food Technology, 6*, 158–165.

Karizaki, V. M., Sahin, S., Sumnu, G., Mosavian, M. T. H., & Luca, A., (2013). Effect of Ultrasound-assisted osmotic dehydration as a pretreatment on deep fat frying of potatoes. *Food and Bioprocess Technology, 6*(12), 3554–3563.

Katekawa, M. E., & Silva, M. A., (2007a). On the influence of glass transition on shrinkage in convective drying of fruits: A case study of banana drying, *Drying Technology, 25*(10), 1659–1666.

Katekawa, M. E., & Silva, M. A., (2007b). Drying rates in shrinking medium: Case study of Banana. *Brazilian Journal of Chemical Engineering, 24*(4), 561–569.

Katsube, T., Tsurunaga, Y., Sugiyama, M., Furuno, T., & Yamasaki, Y., (2009). Effect of air-drying temperature on antioxidant capacity and stability of polyphenolic compounds in mulberry (*Morus alba* L.) leaves. *Food Chemistry, 113*, 964–969.

Kaymak-Ertekin, F., & Gedik, A., (2005). Kinetic modeling of quality deterioration in onions during drying and storage. *Journal of Food Engineering, 68*, 443–453.

Kendall, P., Di Persio, P., & Sofos, J., (2004). *Drying Vegetables: Food and Nutrition Series.* URL: http://nchfp.uga.edu/how/dry/csu_dry_vegetables.pdf (Accessed on 8 November 2019).

Khan, A., & Rahman, M., (2008). Antibacterial, antifungal and cytotoxic activity of amblyone isolated from A. campanulatus. *Indian Journal of Pharmacology, 40*, 41–44.

Khan, A., (1990). Post-harvest losses during processing and preservation of fruits and vegetables. *PhD Dissertation* (pp. 56–123). Institute of Chemistry, University of Punjab, Lahore, Pakistan.

Khattab, N. M., & Barakat, M. H., (2002). Modeling the design and performance characteristics of solar steam-jet cooling for comfort air conditioning. *Solar Energy, 73*, 257–267.

Khetarpal, A., & Kochar, G. K., (2011). *Nutritional Quality of Fruits and Vegetables and their Importance in Human Health*. URL: http://www.ru.org/index.php/health/341-nutritional-quality-of-fruits-and-vegetables-and-their-importance-in-human-health (Accessed on 8 November 2019).

Khraisheh, M. A. M., McMinn, W. A. M., & Magee, T. R. A., (2004). Quality and structural changes in starchy foods during microwave and convective drying. *Food Research International, 37*, 497–503.

Kimambo, C., (2007). *Solar Thermal Applications in Tanzania*. PREA Workshop, Dar-es-Salaam, Tanzania.

Kinabo, J., Mnkeni, A., Nyaruhucha, C. N. M., & Ishengoma, J., (2004). Nutrients content of food commonly consumed in Iringa and Morogoro regions. *Proceedings of the Second Collaborative Research workshop on Food Security* (p. 157). Morogoro. Tanzania.

Klieber, A., (2000). *Chilly Spice Production in Australia* (p. 71). A report for the rural industries research and development corporation. Publication No. 00/33, Project No. UA-38A. Australia.

Kresic, G., Lelas, V., & Simundi, B., (2004). Effects of processing on nutritional composition and quality evaluation of candied celeriac. *Sadhan, 29*(1), 1–12.

Krokida, M. K., Tsami, E., & Maroulis, Z. B., (1998). Kinetics on color changes during drying of some fruits and vegetables. *Drying Technology, 16*(3–5), 667–685.

Kumar, A., & Tiwari, G. N., (2007). Effect of mass on convective mass transfer coefficient during open sun and greenhouse drying of onion flakes. *Journal of Food Engineering, 79*(4), 1337–1350.

Kumar, C., Karim, M. A., & Joardder, M. U. H., (2014). Intermittent drying of food products: A critical review. *Journal of Food Engineering, 121*, 48–57.

Kumar, V., Sinha, A. K., H. Makkar, P. S., & Becker, K., (2010). Dietary roles of phytate and phytase in human nutrition: A review. *Food Chemistry, 120*(4), 945–959.

Kutoš, T., Golob, T., Kač, M., & Plestenjak, A., (2003). Dietary fiber content of dry and processed beans. *Food Chemistry, 80*(2), 231–235.

Langrish, T. A. G., (2008). Characteristic drying curves for cellulosic fibers. *Chemical Engineering Journal, 137*(3), 677–680.

Lee, J., Hwang, Y. S., Kang, I. K., & Choung, M. G., (2015). Lipophilic pigments differentially respond to drying methods in tea (*Camellia sinensis* L.) leaves. *LWT—Food Science and Technology, 61*(1), 201–208.

Lee, S. C., Prosky, L., & DeVries, J. W., (1992). Determination of total, soluble, and insoluble dietary fiber in foods-enzymatic-gravimetric method, MES-TRIS buffer: collaborative study. *Journal of AOAC International, 75*, 395–416.

Lee, S. K., & Kader, A. A., (2000). Preharvest and postharvest factors influencing vitamin C content of horticultural crops. *Postharvest Biology and Technology, 20*, 207–220.

Leong, S. Y., & Oey, I., (2012). Effects of processing on anthocyanins, carotenoids and vitamin C in summer fruits and vegetables. *Food Chemistry, 133*, 1577–1587.

Lester, G. E., (2006). Environmental regulation of human health nutrients (ascorbic acid, carotene, and folic acid) in fruits and vegetables. *Hort Science, 41*, 59–64.

Lestienne, I., Icard-Vernière, C., Mouquet, C., Picq, C., & Trèche, S., (2005). Effects of soaking whole cereal and legume seeds on iron, zinc and phytate contents. *Food Chemistry 89*(3), 421–425.

Lewicki, P. P., (1998a). Some remarks on rehydration of dried foods. *Journal of Food Engineering, 36*(1), 81–87.

Lewicki, P. P., (1998b). Effect of pre drying, drying and rehydration on plant tissue properties: A review. *International Journal of Food Properties, 1*, 1–22.

Lewicki, P. P., (2006). Design of hot air drying for better foods. *Trends in Food Science and Technology, 17*, 153–163.

Lewicki, P. P., Witrowa-Rajchert, D., & Mariak, J., (1997). Changes of structure during rehydration of dried apples. *Journal of Food Engineering, 32*(4), 347–350.

Lijuan, Z., Jianguo, L., Yongkang, P., Guohua, C., & Mujumdar, A. S., (2005). Thermal dehydration methods for fruits and vegetables. *Drying Technology, 23*(9–11), 2249–2260.

Lima, G. P. P., Lopes, T. D. V. C., Rossetto, M. R. M., & Vianello, F., (2009). Nutritional composition, phenolic compounds, nitrate content in eatable vegetables obtained by conventional and certified organic grown culture subject to thermal treatment. *International Journal of Food Science and Technology, 44*(6), 1118–1124.

Lin, T. M., Durance, T. D., & Scaman, C. H., (1998). Characterization of vacuum microwave, air and freeze-dried carrot slices. *Food Research International, 31*, 111–117.

Lintas, C., (1992). Nutritional aspects of fruit and vegetable consumption. In: Lauret, F., (ed.), *Les fruits et légumes dans les économies méditerranéennes: Actes du colloque de Chania* (pp. 79–87). Montpellier, CIHEAM 19.

Liu, R. H., (2003). Health benefits of fruits and vegetables are from additive and synergistic combinations of phytochemicals. *American Journal of Clinical Nutrition, 78*, 517S–520S.

Madrau, M. A., Piscopo, A., Sanguinetti, A. M., Caro, A. D., Poiana, M., & Romeo, F. V., (2008). Effect of drying temperature on polyphenolic content and antioxidant activity of apricots. *European Food: Research and Technology, 228*, 441–448.

Maeda, E. E., & Salunkhe, D. K., (1981). Retention of ascorbic acid and total carotene in solar dried vegetables. *Journal of Food Science, 46,* 1288–1290.

MAFC (Ministry of Agriculture, Food Security and Cooperatives), (2009). *Post-Harvest Technologies for Preparation, Processing and Utilization of Fruits and Vegetables* (p. 89). Makanza Creative Art (MCA), Ministry of Agriculture, Food Security and Cooperatives, Dar-es-Salaam, Tanzania.

Mahan, L. K., & Escott-Stump, S., (2000). *Krause's Food, Nutrition, and Diet Therapy* (10th edn., p. 100). W.B. Saunders Philadelphia, USA.

Mahgoub, S. E. O., & Elhag, S. A., (1998). Effect of milling, soaking, malting, heat-treatment and fermentation on phytate level of four Sudanese sorghum cultivars. *Food Chemistry, 61*(1&2), 77–80.

Manzocco, L., Calligaris, S., Mastrocola, D., Nicoli, M. C., & Lerici, C. R., (2001). Review of non-enzymatic browning and antioxidant capacity in processed foods. *Trends Food Science and Technology, 11*, 340–346.

Mao, L. C., Yang, J., Chen, J. F., & Yu-Ying, Z. Y. Y., (2010). Effects of drying processes on the antioxidant properties in sweet potatoes. *Journal of Agricultural Sciences in China, 9*, 1522–1529.

Marfil, P. H. M., Santos, E. M., & Telis, V. R. N., (2008). Ascorbic acid degradation kinetics in tomatoes at different drying conditions. *LWT-Food Science and Technology*, *41*, 1642–1647.

Marques, L. G., Prado, M. M., & Freire, J. T., (2009). Rehydration characteristics of freeze-dried tropical fruits. *LWT—Food Science and Technology, 42*(7), 1232–1237.

Marshall, M. R., Kim, J., & Wei, C., (2000). Enzymatic browning in fruits, vegetables and sea foods. *Food and Agricultural Organization, 41*, 259–312.

Maskan, A., Kaya, S., & Maskan, M., (2002). Hot air and sun drying of grape leather (pestil). *Journal of Food Engineering, 54*, 81–88.

Maskan, M., (2001). Drying, shrinkage and rehydration characteristics of kiwifruits during hot air and microwave drying. *Journal of Food Engineering, 48*(2), 177–182.

Mason, J. B., Lotfi, M., Dalmiya, N., Sethuraman, K., Deitchler, M., Geibel, S., Gillenwater, K., Gilman, A., Mason, K., & Mock, N., (2001). *The Micronutrient Report: Current Progress in the Control of Vitamin A, Iodine, and Iron Deficiencies*. Ottowa, Micronutrient Initiative/International Development Research Center. URL: http://www.micronutrient.org/frame_HTML/resource_text/publicati ons/mn_report.pdf (Accessed on 8 November 2019).

May, B. K., Sinclair, A. J., Halmos, A. L., & Tran, V. N., (1999). Quantitative analysis of drying behavior of fruits and vegetables. *Drying Technology, 17*(7&8), 1441–1448.

Mayor, L., & Sereno, A. M., (2004). Modeling shrinkage during convective drying of food materials: A review. *Journal of Food Engineering, 61*, 373–386.

McKinley Health Center, (2008). *Macronutrients: The Importance of Carbohydrate, Protein, and Fat*. University of Illinois, USA. URL: http://www.mckinley.illinois.edu/handouts/pdfs/macronutrients.pdf (Accessed on 8 November 2019).

Mepba, H. D., Eboh, L., & Banigo, D. E. B., (2007). Effects of processing treatments on the nutritive composition and consumer acceptance of some Nigerian edible leafy vegetables. *African Journal of Food Agriculture Nutrition and Development, 7*(1), 1–18.

Methakhup, S. M., (2003). Effects of drying methods and conditions on drying kinetics and quality of Indian gooseberry. *M.Sc. Dissertation* (p. 96). Department of Food Engineering, King Mongkut's University of Technology Thonburi, Bangkok, Thailand.

Methakhup, S., Chiewchan, N., & Devahastin, S., (2005). Effects of drying methods and conditions on drying kinetics and quality of Indian gooseberry flake. *LWT- Food Science and Technology, 38*(6), 579–587.

Miao, S., & Roos, Y. H., (2006). Isothermal study of nonenzymatic browning kinetics in spray-dried and freeze-dried systems at different relative vapor pressure environments. *Innovative Food Science and Emerging Technologies, 7*, 182–194.

Mitra, J., Shrivastava, S. L., & Rao, P. S., (2011). Vacuum dehydration kinetics of onion slices. *Food and Bioproducts Processing, 89*, 1–9.

Mokhtarian, M., Tavakolipour, H., & Kalbasi, A. A., (2017). Effects of solar drying along with air recycling system on physicochemical and sensory properties of dehydrated pistachio nuts. *LWT—Food Science and Technology, 75*, 202–209.

Morris, A., Barnett, A., & Burrows, O., (2004). Effect of processing on nutrient content of foods. *Cajarticles, 37*(3), 160–164.

Mosha, T. C., Pace, R. D., Adeyeye, S., Laswai, H. S., & Mtebe, K., (1997). Effect of traditional processing practices on the content of total carotenoid, β-carotene, α-carotene and vitamin A activity of selected Tanzanian vegetables. *Plant Foods for Human Nutrition, 50,* 189–201.

Mota, C. L., Luciano, C., Dias, A., Barroca, M. J., & Guine, R. P. F., (2010). Convective drying of onion: kinetics and nutritional evaluation. *Food and Bioproducts Processing, 88,* 115–123.

Mujumdar, A. S., & Huang, L. X., (2007). Global R&D needs in drying. *Drying Technology, 25*(4), 647–658.

Mujumdar, A. S., (1991). Drying technologies of the future. *Drying Technology, 9,* 325–347.

Mulokozi, G., & Svanberg, U., (2003). Effect of traditional open sun-drying and solar cabinet drying on carotene content and vitamin A activity of green leafy vegetables. *Plant Foods for Human Nutrition, 58,* 1–15.

Munde, A. V., Agrawal, Y. C., & Shrikande, V. J., (1988). Process development for multistage dehydration of onion flakes. *Journal of Agricultural Engineering, 25*(1), 19–24.

Musa, J. J., Idah, P. A., & Olaleye, S. T., (2010). Effect of temperature and drying time on some nutritional quality parameters of dried tomatoes. *Assumption University: AU Journal of Technology, 14*(1), 25–32.

Ndawula, J., Kabasa, J. D., & Byaruhanga, Y. B., (2004). Alterations in fruit and vegetable b-carotene and vitamin C content caused by open-sun drying, visqueen-covered and polyethylene-covered solar-dryers. *African Health Sciences, 4,* 125–130.

Oberoi, D. P. S., & Sogi, D. S., (2015). Effect of drying methods and maltodextrin concentration on pigment content of watermelon juice powder. *Journal of Food Engineering, 165,* 172–178.

Obied, H. K., Bedgood, D. R., Prenzier, P. D., & Robards, K., (2008). Effect of processing conditions, pre-storage treatment, and storage conditions on the phenol content and antioxidant activity of olive mill waste. *Journal of Agricultural and Food Chemistry, 56,* 3925–3932.

Ottaway, P. B., (1993). The technology of vitamins in food. Published by Blackie Academic and Professional, an important of Chapman & Hall. In: Mujumdar, A. S., (ed.), *Dehydration of Products of Biological Origin (2004)* (p. 51). Science Publisher, Inc. Enfield, USA.

Özkan, I. A., Akbudak, B., & Akbudak, N., (2007). Microwave drying characteristics of spinach. *Journal of Food Engineering, 78,* 577–583.

Park, K. H., Park, Y. D., Han, J. M., Lin, K. R., Lee, B. W., & Jeong, I. Y., (2006). Antiatherosclerotic and anti-inflammatory activities of catecholic xanthones and flavonoids isolated from *Cudrania tricuspidata. Bioorganic and Medicinal Chemistry Letters, 16,* 5580–5583.

Perera, C. O., & Baldwin, E. A., (2001). Biochemistry of fruits and its implications on processing. In: Arthey, D., & Ashurst, R. P., (eds.), *Fruit Processing: Nutrition, Products, and Quality Management* (2nd edn., pp. 1–123). Aspen Publishers, Inc., New York, USA.

Perera, C. O., (2005). Selected quality attributed of dried foods. *Journal of Drying Technology, 23*(4), 717–730.

Perkins-Veazie, P., & Collins, J. K., (2001). Contributions of non-volatile phytochemicals to nutrition and flavor. *Horticulture Technology, 11,* 539–546.

Perumal, R., (2007). Comparative performance of solar cabinet, vacuum assisted solar and open sun drying methods. *M.Sc. Dissertation* (p. 110), Department of Bioresource Engineering, McGill University, Montreal, Canada.

Phillippy, B. Q., & Lin, R. M., B., (2004). Analysis of phytate in raw and cooked potatoes. *Journal of Food Composition and Analysis, 17*(2), 217–226.

Piga, A., Catzeddu, P., Farris, S., Roggio, T., Sanguinetti, A., & Scano, E., (2005). Texture evaluation of "Amaretti" cookies during storage. *Food Research and Technology, 221*, 387–391.

Podsedek, A., (2007). Natural antioxidants and antioxidant capacity of Brassica vegetables: A review. *LWT-Food Science and Technology, 40*, 1–11.

Poll, L., & Thi, A. D. P., (2004). Aroma and vitamin changes in processed fruits. *Journal of Agriculture and Food Chemistry, 51*, 19–23.

Popov, O. A., Efron, B. G., & Milakova, E., (1976). *Vse-soyuznyi Nauchno-issledovatel' shii Institut Konservnoii Ovoshchesushill'noi Promyshlenn-osti., 24*, 43–50.

Pu, Y. Y., & Sun, D. W., (2017). Combined hot-air and microwave-vacuum drying for improving drying uniformity of mango slices based on hyperspectral imaging visualization of moisture content distribution. *Biosystems Engineering, 156*, 108–119.

Quintero-Ramos, A., Bourne, M. C., & Anzaldua-Morales, A., (1992). Texture and rehydration of carrots as affected by low temperature blanching. *Journal of Food Science, 57*, 1127–1137.

Rahman, M. M., Khan, M. M. R., & Hosain, M. M., (2007). Analysis of vitamin C (ascorbic acid) contents in various fruits and vegetables by UV-spectrophotometry. *Bangladesh Journal of Scientific and Industrial Research, 42*(4), 417–424.

Rahman, M. S., (1995). *Food Properties Handbook* (p. 528). CRC Press, New York, USA.

Raice, R. T., Chiau, E., Sjoholm, I., & Bergenståhl, B., (2015). The loss of aroma components of the fruit of *Vangueria infausta* L. (African medlar) after convective drying. *Drying Technology, 33*(8), 887–895.

Rathnayaka, M. C. M., Karunasena, H. C. P., Gu, Y. T., Guan, L., & Senadeera, W., (2017). Novel trends in numerical modeling of plant food tissues and their morphological changes during drying: A review. *Journal of Food Engineering, 194*, 24–39.

Ratti, C., (1994). Shrinkage during drying of foodstuffs. *Journal of Food Engineering, 23*, 91–95.

Ratti, C., (2001). Hot air and freeze-drying of high-value foods: A review. *Journal of Food Engineering, 49*(4), 311–319.

Ravindran, V., Ravindran, G., & Sivalogan, S., (1994). Total and phytate phosphorus contents of various foods and feedstuffs of plant origin. *Food Chemistry, 50*(2), 133–136.

Ribeiro, S. M. R., Queiroz, J. H., Queiroz, M. E. L. R., Campos, M. F., & Santana, P. M. H., (2007). Antioxidant in mango (*Mangifera indica* L.) pulp. *Plant Foods for Human Nutrition, 62*, 13–17.

Ricce, C., Rojas, M. L., Miano, A. C., Siche, R., & Augusto, P. E. D., (2016). Ultrasound pre-treatment enhances the carrot drying and rehydration. *Food Research International, 89*(Part 1), 701–708.

Richard, A. H., & Ferrier, D. R., (2011). *Lippincott's Illustrated Reviews: Biochemistry.* (5th edn., pp. 43–52). Fibrous Protein, Lippincott Williams and Wilkins, New York, USA.

Roig, M. G., Bello, J. F., Rivera, Z. S., & Kennedy, J. F., (1999). Studies on the occurrence of non-enzyamtic browning during storage of citrus juice. *Food Research International, 32*, 609–619.

Sablani, S. S., (2006). Drying of fruits and vegetables: retention of nutritional/ functional quality. *Drying Technology, 24*, 428–432.

Sagar, V. R., & Suresh, K. P., (2010). Recent advances in drying and dehydration of fruits and vegetables: A review. *Journal of Food Science and Technology, 47*(1), 15–26.

Sahin, S., & Sumnu, S. G., (2006). *Physical Properties of Foods*. Springer.

Salunkhe, D. K., (1974). *Storage, Processing and Nutritional Quality of Fruits and Vegetables* (pp. 29, 30). CRC Press, New York, USA.

Santos-Sánchez, N. F., Valadez-Blanco, R., Gómez-Gómez, M. S., Pérez-Herrera, A., & Salas-Coronado, R., (2012). Effect of rotating tray drying on antioxidant components, color and rehydration ratio of tomato saladette slices. *LWT—Food Science and Technology, 46*(1), 298–304.

Segura-Carretero, A., Garcia-Salas, P., Morales-Soto, A., & Fernández-Gutiérrez, A., (2010). Phenolic-compound-extraction systems for fruit and vegetable samples. *Molecules 15*, 8813–8826.

Seremet, L., Botez, E., Nistor, O. V., & Andronoiu, D. G. G. D., (2016). Effect of different drying methods on moisture ratio and rehydration of pumpkin slices. *Food Chemistry, 195*, 104–109.

Shahidi, F., & Naczk, M., (2004). Extraction and analysis of phenolics in food. *Journal of Chromatograph, 1054*(1&2), 95–111.

Shrimpton, D. H., (1993). Nutritional aspects of vitamins. In: Ottaway, O. P., (ed.), *The Technology of Vitamins in Food* (pp. 22–25). Kluwer Academic Publishers, USA.

Sobukola, O. P., Dairo, O. U., Sanni, L. O., Odunewu, A. V., & Fafiolu, B. O., (2007). Thin layer drying process of some leafy vegetables under open sun. *Journal of Food Science and Technology International, 2*, 13–35.

Somogyi, L. P., & Luh, B. S., (1986). Dehydration of fruits. In: Woodroof, J. G., & Luh, B. S., (eds.), *Commercial Fruit Processing* (2nd edn., pp. 353–405). Westport: AVI Publishing Co., Inc.

Sonni, A., (2002). *Importance of Minerals and Trace Minerals in Human Nutrition*. URL: http://www.mgwater.com/import.shtml (Accessed on 8 November 2019).

Spiller, G. A., (2001). Dietary fiber in prevention and treatment of disease. In: Spiller, G. A., (ed.), *Handbook of Dietary Fiber in Human Nutrition* (pp. 363–431). CRC Press, New York, USA.

Steinke, J. A., Frick, C. M., Gallagher, J. A., & Strassburger, K. J., (1989). Influence of Microwave heating on flavor. In: Parliment, T., McGorrin, R., & Ho, C. T., (eds.), *Thermal Generation of Aromas* (pp. 520–525). ACS Symposium Series, No. 409, American Chemical Society, Washington, DC, USA.

Suvarnakuta, P., Chaweerungrat, C., & Devahastin, S., (2011). Effects of drying methods on assay and antioxidant activity of xanthones in mangosteen rind. *Food Chemistry, 125*, 240–247.

Swasdisevi, T., Soponronnarit, S., Prachayawarakorn, S., & Phetdasada, W., (1999). Drying of chopped spring onion using fluidization technique. *Drying Technology, 17*(6), 1191–1199.

Talla, A., Puiggali, J. R., Jomaa, W., & Jannot, Y., (2004). Shrinkage and density evolution during drying of tropical fruits: Application to banana. *Journal of Food Engineering, 64*, 103.

Tang, Y. C., & Chen, B. H., (2000). Pigment change of freeze-dried carotenoid powder during storage. *Food Chemistry, 69*(1), 11–17.

Temu, A., Chove, B., & Ndabikunze, B., (2008). *Development of Enterprise in Solar Drying of Fruit and Vegetables for Employment Creation* (p. 15). Communication workshop for DANIDA pilot projects in SUA. SUA, Morogoro, Tanzania.

Toğrul, İ. T., & Pehlivan, D., (2004). Modeling of thin layer drying kinetics of some fruits under open-air sun drying process. *Journal of Food Engineering, 5*(3), 413–425.

Traub, D. A., (2002). *The Drying Curve Part 2*. Applying the drying curve to your drying process. URL: http://www.processheating.com/ext/resources/PH/Home/Files/PDFs/1002ph_dryingfiles.pdf (Accessed on 8 November 2019).

Uhlig, B. A., & Walker, R. R., (1996). Mode of action of the drying emulsion used in dried vine fruit production II. The effect of emulsion pH. *Australian Journal of Grape and Wine Research, 2*, 91–96.

UNCTAD, (2007). *Safety and Quality of Fresh Fruit and Vegetables: A Training Manual for Trainers* (p. 140). United Nations, New York and Geneva.

Van Ruth, S. M., & Roozen, J. P., (1994). Gas chromatography/sniffing port analysis and sensory evaluation of commercially dried bell peppers (Capsicum annuum) after rehydration. *Food Chemistry, 51*(2), 165–170.

Vega-Galvez, A., Di Scala, K., Rodriguez, K., Lemus-Mondaca, R., Miranda, M., Lopez, J., & Perez-Won, M., (2009). Effect of air-drying temperature on physico-chemical properties, antioxidant capacity, color and total phenolic content of red pepper (*Capsicum annuum* L. var. Hungarian). *Food Chemistry, 117*, 647–653.

Vergeldt, F. J., Van Dalen, G., Duijster, A. J., Voda, A., Khalloufi, S., Van Vliet, L. J., Van As, H., Van Duynhoven, J. P. M., & Van Der Sman, R. G. M., (2014). Rehydration kinetics of freeze-dried carrots. *Innovative Food Science and Emerging Technologies, 24*, 40–47.

Verma, D. K., Kapri, M., Billoria, S., Mahato, D. K., & Srivastav, P. P., (2017). Effects of thermal processing on nutritional composition of green leafy vegetables: A review. In: Verma, D. K., & Goyal, M. R., (eds.), *Engineering Interventions in Foods and Plants: Series of Innovations in Agricultural and Biological Engineering*. Apple Academic Press, USA.

Voda, A., Homan, N., Witek, M., Duijster, A., Van Dalen, G., Van Der Sman, R., Nijsse, J., Van Vliet, L., Van As, H., & Van Duynhoven, J., (2012). The impact of freeze-drying on microstructure and rehydration properties of carrot. *Food Research International, 49*(2), 687–693.

Waje, S. S., Meshram, M. W., Chaudhary, V., Pandey, R., Mahanawar, P. A., & Thorat, B. N., (2005). Drying and shrinkage of polymer gels. *Brazilian Journal of Chemical Engineering, 22*(2), 209–215.

Wakayama, T., (1995). Polyphenol oxidase activity in Japanese apples. In: Lee, C. Y., & Whitaker, J. R., (eds.), *Enzymatic Browning and its Prevention* (pp. 251–266). American Chemical Society, Washington DC, USA.

Wedzicha, B. L., (1984). *Chemistry of Sulphur Dioxide in Foods* (p. 111). Elsevier Applied Science, New York, USA.

Whitaker, J. R., & Lee, C. Y., (1995). Recent advances in chemistry of enzymatic browning. In: Lee, C. Y., & Whitaker, R. J., (eds.), *Enzymatic Browning and its Prevention* (pp. 2–7). American Chemistry Society Symposium, Series. 600, Washington DC, USA.

Wiriya, P., Paiboon, T., & Somchart, S., (2009). Effect of drying air temperature and chemical pretreatments on quality of dried chili. *International Food Research Journal, 16,* 4–7.

Yang, C. S. T., & Atallah, W. A., (1985). Effect of four drying methods on the quality of intermediate moisture low bush blueberries. *Journal of Food Science, 50*(5), 1233–1237.

Yaylayan, V. A., & Roberts, D. D., (2001). Generation and release of food aromas under microwave heating. In: Datta, A. K., & Anantheswaran, R. C., (eds.), *Handbook of Microwave Technology for Food Applications* (pp. 173–189). Marcel Dekker, New York.

Yeung, D. L., & Laquatra, I., (2003). In: Heinz, H. J., (ed.), *Heinz Handbook of Nutrition* (9th edn., p. 266). Company, USA.

Yuan, J. P., & Chen, F., (1998). Degradation of ascorbic acid in aqueous solution. *J. Agric. Food Chem., 46,* 5078–5082.

Zhang, L. W., Ji, H. F., Du, A. L., Xu, C. Y., Yang, M. D., & Li, F. F., (2012). Effects of drying methods on antioxidant properties in *Robinia pseudoacacia* L. flowers *Journal of Medicinal Plants Research, 6*(16), 3233–3239.

Zielinska, M., & Markowski, M., (2016). The influence of microwave-assisted drying techniques on the rehydration behavior of blueberries (*Vaccinium corymbosum* L.). *Food Chemistry, 196,* 1188–1196.

Zogzas, N. P., Maroulis, Z. B., & Marinos-Kouris, D., (1994). Densities, shrinkage and porosity of some vegetables during air drying. *Drying Technology, 12*(7), 1653–1666.

# CHAPTER 4

# Disinfection of Drinking Water by Low Electric Field

ASAAD REHMAN SAEED AL-HILPHY, NAWFAL A. ALHELFI, and
SAHER SABIH GEORGE

*Department of Food Science, College of Agriculture, University of Basrah,
Basra City, Iraq, E-mail: aalhilphy@yahoo.co.uk (A.R.S. Al-Hilphy);
E-mail: nawfalalhelfi@gmail.com (N.A. Alhelfi);
E-mail: Saher_sg@yahoo.com (S.S. George)*

## 4.1 INTRODUCTION

Water is one of very important elements for human life because the life was begun *via* water only on the earth (Jequier and Constant, 2010; Popkin et al., 2010; Dkhar et al., 2014). Three-quarters of the earth is covered by the water. There are two types of water in drylands, surface water, and groundwater. Drinking water is beneficial for humans, not harmful (Phillips et al., 1984; Fadda et al., 2012). On the other hand, drinking water does not contain pathogenic microbes and other microbial or chemical contaminants (Hammer, 1975; Phillips et al., 1984; Fadda et al., 2012). Contaminated drinking water is a source of a lot of diseases such as bacillary dysentery, typhoid, poliomyelitis, cholera, and bilharziasis (Fawell and Chipman, 2000; Fawell and Standfield, 2001; Fawell and Nieuwenhuijsen, 2003; EPA, 2008; USCB, 2008). Disinfection is a treatment used for eliminating all microorganisms from water (Charles et al., 2002). There are two major means used for disinfecting water, i.e., the physical and chemical ways. Physical disinfection includes: rising water temperature over 75°C for a short time firstly, and secondly, using ultraviolet (UV) radiation ranged between 240–280 nm. UV radiation (UVR) assaults deoxyribonucleic acid (DNA) of bacteria directly (Rastogi et al., 2010), and bacteria deprives its capability on the proliferation, then it destroys and dies after that. The third type of physical disinfection is the filtration by making use of sand filters to eliminate a huge percentage of microorganisms (about 90%). Besides these, the reverse osmosis (RO) technology consisting of cellulose membranes with 0.001 μm diameter pores are used for disinfection drinking water (Wimalawansa, 2013; Lee et al., 2016). Machado et al. (2011) narrated that the moderate electrical field at 220 V cm$^{-1}$ destroyed *E. coli* by 3 log cycles. Chemical ways are widely used for sterilizing drinking water like ClO (dioxide chlorine) and chlorine (Trakhtman, 1946; Symons et al., 1977; Sussman, 1978). Al-Manhel et al. (2016) stated that the chitosan (nonthermal treatment) is used to disinfect drinking water, and eliminated total bacteria, fecal coliform bacteria, staphylococci, and total coliform bacteria as indicated from the microbiological analysis.

## 4.2  NONTHERMAL STERILIZER OF WATER

The nonthermal sterilization apparatus of water comprises of 5 liters capacity stainless steel tank and four electrodes made of stainless steel. Their diameter and length are 0.4 cm and 2 cm, respectively, and the distance between each other is 1 cm. The other components are handle voltage regulator (voltage range between 0 to 230 V) and valve, as illustrated in Figure 4.1. The mass flow rate of water is 0.08 kgs$^{-1}$ (Al-Hilphy et al., 2016).

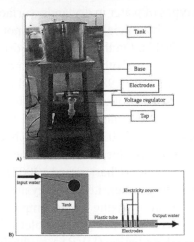

**FIGURE 4.1**  Nonthermal sterilizer by a low electrical field. A) A laboratory set up and B) Schematic view.
(*Source*: Reprinted from Al-Hilphy, Alhelfi, and George, (2016)).

### 4.2.1  ELECTRICAL CONDUCTIVITY (EC)

Electrical conductivity (EC) indicates how water transports the electric charge, as well as the effect of the water chemical composition on the EC. Anderson (2008) stated that when water ions like sodium chloride (NaCl) increases, the EC increases. EC is calculated by equation (1) (Wang and Sastry, 1993; Icier et al., 2008):

$$\sigma = \frac{Id}{VA} \tag{1}$$

where, $\sigma$, $A$, $d$, and $V$ are an EC (Sm$^{-1}$), section area (m$^2$), the distance between electrodes (cm), and voltage (V), respectively.

Figure 4.2 illustrates the relation between the EC (Sm$^{-1}$) and the electrical field (Vcm$^{-1}$), and the current (A). EC is not significantly decreased with the excess of the electric field, but the slight effect is attributed to the voltage that increased with the rise of the electric field that led to reduce the EC, as illustrated in the Equation (2).

$$\sigma = \frac{(I)}{E_f A} \tag{2}$$

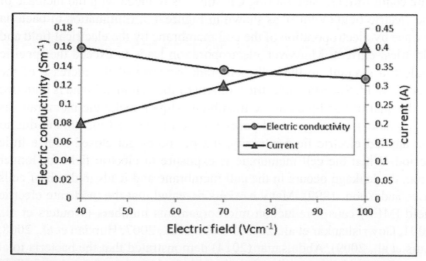

**FIGURE 4.2**   The relation between water electrical conductivity and electric field. (*Source*: Reprinted from Al-Hilphy, Alhelfi, and George, (2016)).

EC has an important role in inactivation of microorganisms (Machado et al., 2011). Castro et al. (2004) had noticed that the differences between strawberry-apple sauce EC and strawberry at various electrical field values were not significant at a temperature between 22–100°C, but they found that the effect of electric field on the strawberry filling EC was significant. The differences in the electric conductivities with the changing electric field depend on the quantity of salts, current, voltage, temperature, food properties, and pH. Castro et al. (2003) demonstrated that the EC decreases with the increase of the electric field due to available of air bubbles when the electric field increases. The air bubbles increase with the increase of EC (Icier and Ilicali, 2005).

The results also show that the EC values are ranging between 0.127–0.159 Sm⁻¹, also the electric current vary between 0.20–0.40 Å. There are many published papers that studied the relation between the EC and electric field like Castro et al. (2004) for strawberry-apple sauce, Icier et al. (2008) for milk, Mohsin (2001) for milk, Abdulsattar (2014) for milk and Al-Hilphy et al. (2015) for wheat bran.

### 4.2.2   TOTAL BACTERIA COUNT

The count of total bacteria (log CFUml⁻¹) is reduced with the increase of the electric field (Vcm⁻¹), as shown in Figure 4.3. Elimination of bacteria is done by electroporation of the cell membrane by the electrical field and electrical current. Moreover, electroporation leads to exceed the specific dielectric strength in the cell membrane because of the increase of the electric field. Specific dielectric strength of the cell membrane is bounded to the quantity of lipids in the membrane (lipids make the membrane as an insulator) forming varied pores in size on the cell membrane due to the applied electric field intensity, but the pores get closed after a little period. When the cell membrane is exposure to electric field for longer time, the leakage occurs in the cell membrane and leads to death of cells (Lee and Yoon, 1999). Many workers revealed that the moderate electric field (MEF) causes reduction microorganisms numbers (Wouters et al., 2001; Gowrishankar et al., 2005; Pereira et al., 2007; Huixian et al., 2008; Luis et al., 2009). Abdulsattar (2014) demonstrated that the bacteria total count in milk decreases significantly with the increase of the electric field intensity, and reached to zero. The relationship between electric field and log bacteria total count is exponential, and $R^2 = 0.9938$ as illustrated in Figure 4.3 and the following equation:

$$Log(CFU\ ml^{-1}) = 6.4255e^{-0.017Ef} \qquad (3)$$

where, $Ef$ is the electrical field ($Vcm^{-1}$).

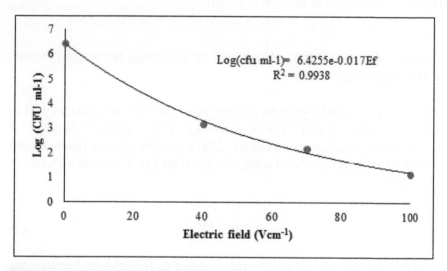

**FIGURE 4.3**  Bacteria total count vs. electrical field.
(*Source*: Abdulsattar (2014)).

### 4.2.3  TOTAL COLIFORM BACTERIA AND E. COLI

Table 4.1 illustrates impact of electric field on the *E. coli* and total coliform bacteria (CFU ml⁻¹). The *E. coli* and total coliform were $1\times10^1$ CFU ml⁻¹ and $4\times10^2$ CFU ml⁻¹, respectively for control, and then decreased to nil at all electrical field values (40, 70 and 100 Vcm⁻¹). This effect is because of electroporation of bacteria *via* passing electrical current into the water (water became as a resistor) and the electrical field has generated that lead to eliminate the microorganisms. Machado et al. (2011) announced that the alternative current (AC) eliminates *Escherichia coli* at 25°C. In addition, they discovered that the *Escherichia coli* are decreased with the increasing electric field, due to altered cell membrane ranging between 0.1–50 µm.

**TABLE 4.1** Effect of Electrical Field on the *E. coli* and Total Coliform Bacteria

| Electric Field (Vcm$^{-1}$) | Total Coliform Bacteria (CFUml$^{-1}$) | *E. coli* (CFUml$^{-1}$) |
|---|---|---|
| Control | $4 \times 10^2$ | 10 |
| 60 | Nil | Nil |
| 80 | Nil | Nil |
| 100 | Nil | Nil |

(*Source*: Reprinted from Al-Hilphy et al. (2016)).

### 4.2.3.1 MATHEMATICAL MODELING OF SURVIVAL MICROORGANISM RATIO TO TOTAL

Mathematical modeling can be done by solver in excel program, as well as solving equations 4 and 5 can be solved by SPSS (Statistical Package for the Social Science) program (SPSS, 2001). Survival microorganism ratio to total (C/CT) is calculated using the equation (4) (Fernandez-Molina et al., 2001).

$$\left(\frac{C}{C_T}\right) = \exp.\left(kV\frac{t}{Q}\right) \tag{4}$$

The exponential equation is representing to the relation between the time and outlet microorganisms. We will inter the effect of electrical field in equation (4) to destruction of microorganisms by electrical field as shown in equation (5), so the k unit in equation (4) is became 1/Vcm$^{-1}$ instead s$^{-1}$.

$$\left(\frac{C}{C_T}\right) = \exp.\left(\frac{-kVE_f t}{Q}\right) \tag{5}$$

Equation (5) is a novel equation that describes the impact of electrical field on the survival microorganisms' ratio. Where, V is the treatment chamber volume (m$^3$), *Ef* is the electrical field (Vcm$^{-1}$), Q is the discharge (m$^3$/s), k is the constant (1/Vcm$^{-1}$), t is the treatment time (s). The constant k can be predicted using mathematical modeling.

The predicted C/CT and experimental have significantly decreased with increasing electrical field, this may be attributed to the electroporation of microorganisms by the electric field. It can be seen from Figure 4.4 that there is a convergent between predicted and experimental C/CT ratio. Moreover, statistical parameters values have a good fitting between the predicted and experimental data, i.e., the determination coefficient ($R^2$), root mean square error (RMSE), and chi-square ($x^2$) were reached 0.991591, 0.00010503 and $5.52 \times 10^{-9}$, respectively.

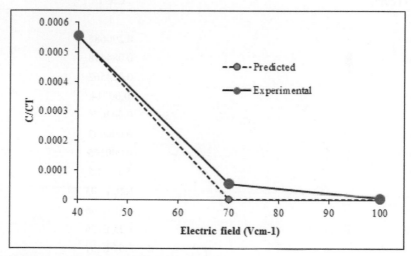

**FIGURE 4.4**   C/CT ratio *vs.* electrical field.
(*Source*: Reprinted from Al-Hilphy, Alhelfi, and George (2016)).

On the other hand, the constant k was 0.1259124 ($1/Vcm^{-1}$) as shown in Table 4.2. And Equation (5) becomes as follow:

$$\left( \frac{C}{C_T} \right) = \exp. \left( -0.1259124 \; \frac{VE_f t}{Q} \right) \tag{6}$$

The proposed equation (6) is a new equation used for describing the impact of the electrical field strength on the C/CT ratio. Results of the equation (6) are illustrated in Table 4.3. When, Q = 0.00008 $m^3 s^{-1}$, V = 0.000157 $m^3$, t = 0.5 s, k = 0.1259124$1/Vcm^{-1}$ and *Ef* values between 0–230 $Vcm^{-1}$.

**TABLE 4.2** Statistical Parameters of the Mathematical Modeling

| k (1/Vcm⁻¹) | RMSE | X² | R² |
|---|---|---|---|
| 0.1259124 | 0.00010503 | $5.52 \times 10^{-9}$ | 0.991591 |

(*Source*: Reprinted from Al-Hilphy et al. (2016)).

**TABLE 4.3** Outcomes of the Equation (9)

| Ef (Vcm⁻¹) | C/CT |
|---|---|
| 0 | 1 |
| 10 | 0.290685 |
| 20 | 0.084498 |
| 30 | 0.024562 |
| 40 | 0.00714 |
| 50 | 0.002075 |
| 60 | 0.000603 |
| 70 | 0.000175 |
| 80 | 5.1 E–05 |
| 90 | 1.48 E–05 |
| 100 | 4.31 E–06 |
| 110 | 1.25 E–06 |
| 120 | 3.64 E–07 |
| 130 | 1.06 E–07 |
| 140 | 3.08 E–08 |
| 150 | 8.94 E–09 |
| 160 | 2.6 E–09 |
| 170 | 7.55 E–10 |
| 180 | 2.2 E–10 |
| 190 | 6.38 E–11 |
| 200 | 1.86 E–11 |
| 210 | 5.39 E–12 |
| 220 | 1.57 E–12 |
| 160 | 2.6 E–09 |
| 230 | 4.56 E–13 |

when, Q = 0.00008 m³s⁻¹, V = 0.000157 m³ and t = 0.5 s.
(*Source*: Reprinted from Al-Hilphy et al. (2016)).

### 4.2.4 RATE OF MICROORGANISM'S DESTRUCTION

Microorganisms' destruction rate is calculated from Equation (7) (Esplugas et al., 2001):

$$r = -kc \tag{7}$$

where, k is constant ($s^{-1}$), r is the microorganisms' destruction rate (CFU/L $s^{-1}$). C is the numbers of microorganisms' ($CFU/L^{-1}$). Figure 4.5 (A) illustrates the relation between the microorganisms' destruction rates (CFU/L $s^{-1}$) and electrical field ($Vcm^{-1}$). The microorganisms' destruction rates are ranged between $-281$ to $-2.81$ CFU/L $s^{-1}$ at range of the electrical field between of 40–100, respectively. The negative sign indicates that the microorganisms reduce. The reduction happens due to the electrical field impact. Reducing microorganism percentage (SR) is calculated from the following equation:

$$SR = N1 - N2 / N1 \tag{8}$$

where, $N$ is the initial microorganisms (CFU $ml^{-1}$), $N$ is the final microorganisms (CFU $ml^{-1}$). Figure 4.5 (B) shows that the percentage of microorganisms' reduction is significantly affected by the electrical field. The reduction is 99.94–99.99% for electrical field range between 40–100 $Vcm^{-1}$.

### 4.3 PROSPECTIVE FUTURE AND RESEARCH OPPORTUNITIES

The use of electrical field as a nonthermal treatment for water is a novel technology. The prospective future of this technology is to produce a large plant for water sterilization by electrical field connected with the RO units. Moreover, this technology needs a lot of deep studies to develop it further. Also, it necessitates use of nanotechnology for manufacturing electrodes.

### 4.4 SUMMARY AND CONCLUSION

A continuous nonthermal disinfection device is used for disinfecting drinking water by low electrical field. It consists of reservoir, four

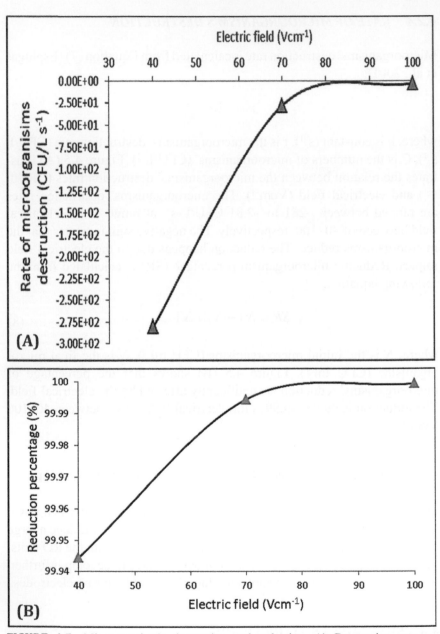

**FIGURE 4.5** Microorganism's destruction and reduction. A) Destruction rate *vs.* electrical field and B) Reduction percentage *vs.* electrical field.
(*Source:* Reprinted from Al-Hilphy, Alhelfi, and George (2016).

stainless steel electrodes and valve. The investigated parameter were EC, count of total aerobic bacteria, total coliform bacteria count, *E. coli*, survival microorganism's ratio to total (C/CT) calculation, microorganisms' destruction rate and decreasing microorganism percentage. Water EC ranged between 0.127–0.159 Sm$^{-1}$. In addition, total bacteria count; total coliform bacteria count, *E. coli* and ratio of survival microorganism's to total were reduced significantly with increasing the electrical field strength. Besides, microorganisms' destruction rate is ranging between –281 to –2.81 CFU/L s$^{-1}$ and the percentage of reducing microorganism ranged between 99.94–99.99% at all electrical field strength (from 40–100 Vcm$^{-1}$), respectively. Moreover, predicted C/CT equation was developed by entering the electrical field strength impact in it.

## KEYWORDS

- **chemical contaminants**
- **chitosan**
- **coliform bacteria**
- **destruction rate**
- **disinfection**
- **drinking water**
- *E. coli*
- **electrical conductivity**
- **fecal coliform bacteria**
- **low electric field**
- **mathematical modeling**
- **osmosis technology**
- **sterilization**
- **UV radiation**

## REFERENCES

Abdulsattar, A. R., (2014). Designing, manufacturing and testing of an apparatus for nonthermal milk pasteurization by electric field. *M.Sc. Thesis* (p. 116). Food Science, Agriculture College, Basrah University, Iraq.

Al-Hilphy, A. R. S., Alhelfi, N. A., & George, S. S., (2016). The continuous nonthermal sterilization of drinking water using low electric field. *Journal of Biology, Agriculture and Healthcare, 6*(14), 83–91.

Al-Hilphy, A. R. S., AlRikabi, A. K. J., & Al-Salim, A. M., (2015). Extraction of phenolic compounds from wheat bran using ohmic heating. *Food Science and Quality Management, 43,* 21–28.

Al-Manhel, A. J., Al-Hilphy, A. R., & Niamah, A. K., (2016). Extraction of chitosan, characterization and its use for water purification. *Journal of the Saudi Society of Agricultural Sciences.* Accepted for publication, doi:10.1016/j.jssas.2016.04.001.

Anderson, D., (2008). Ohmic heating as an alternative food processing technology. *MS Thesis* (p. 45). Kansas state university, Food Science Institute, College of Agriculture. *Manhattan.*

Castro, I., Teixeira, J. A., Salengke, S. Sastary, S. K., & Vicente, A. A., (2004). Ohmic heating of strawberry products: Electrical conductivity measurement and ascorbic acid degradation kinetic. *Innovative Food Science and Emerging Technologies, 5,* 27–36.

Castro, I., Teixeira, J., & Vicente, A., (2003). The influence of field strength, sugar and solid content on electrical conductivity of strawberry products. *Journal of Food Process and Engineering, 26,* 17–29.

Charles, D., Steffen, E. R., & Backer, H., (2002). Water disinfection for international and wilderness travelers. *Clinical Infectious Diseases, 34*(3), 355–364.

Dkhar, E. N., Dkhar, P. S., & Anal, J. M. H., (2014). Trace elements analysis in drinking water of Meghalaya by using graphite furnace-atomic absorption spectroscopy and in relation to environmental and health issues. *Journal of Chemistry*, 1–8. Article ID 975810, http://dx.doi.org/10.1155/2014/975810 (Accessed on 8 November 2019).

EPA (Environmental Protection Agency), (2008). *Factoids: Drinking Water and Ground Water Statistics for 2007.* URL: https://nepis.epa.gov/Exe/ZyPDF.cgi/P100N2VG. PDF?Dockey=P100N2VG.PDF (Accessed on 8 November 2019).

Esplugas, S., Pagan, R., Barbosa-Canovas, G. V., & Swanson, B. G., (2001). Engineering aspects of the continuous treatment of fluid foods by pulsed electric fields. In: Gustavo, V., Barbosa-Canovas, Q., Howard, Z., & Tabilo-Munizaga, G., (eds.), *Pulsed Electric Fields in Food Processing: Fundamental Aspects and Applications.* Taylor & Francis, UK.

Fadda, R., Rapinett, G., Grathwohl, D., Parisi, M., Fanari, R., Calò, C. M., & Schmitt, J., (2012). Effects of drinking supplementary water at school on cognitive performance in children. *Appetite, 59*(3), 730–737.

Fawell, J. K., & Standfield, G., (2001). Drinking water quality and health. In: Harrison, R. M., (ed.), *Pollution: Causes, Effects and Control* (4th edn., pp. 59–80). Royal Society of Chemistry, London.

Fawell, J., & Chipman, K., (2000). Endocrine disrupters, drinking water and public reassurance. *Water Environ Manage, 5,* 4–5.

Fawell, J., & Nieuwenhuijsen, M. J., (2003). Contaminants in drinking water. *British Medical Bulletin, 68*(1), 199–208.

Fernandez-Molina, J. J., Barkstrom, E., Torstensson, P., Barbosa-Canovas, G. V., & Swanson, B. G., (2001). Inactivation of *Listeria innocua* and Pseudomonas fluorescens in skim milk treated with Pulsed Electric Fields (PEF). In: Gustavo, V., Barbosa-Canovas, Q.,

Howard, Z., & Tabilo-Munizaga, G., (eds.), *Pulsed Electric Fields in Food Processing: Fundamental Aspects and Applications*. Taylor & Francis, UK.

Gowrishankar, T. R., Stewart, D. A., & Weaver, J. C., (2005). Model of a confined spherical cell in uniform and heterogeneous applied electric fields. *Bioelectrochemistry, 68*, 185–194.

Hammer, M. J., (1975). *Water and Waste-Water Technology*. New York: John Wiley & Sons, USA.

Huixian, S., Shuso, K., Jun-ichi, H., Kazuhiko, I., Tatsuhiko, W., & Toshinori, K., (2008). Effect of ohmic heating on microbial counts and denaturation of protein in milk. *Food Science and Technology Research, 14*(2), 117–123.

Icier, F., & Ilicali, C., (2005). The effects of concentration on electrical conductivity of orange juice concentrates during ohmic heating. *European Food Research and Technology, 220*, 406–414.

Icier, F., Yildiz, H., & Baysal, T., (2008). Polyphenoloxidase deactivation kinetics during Ohmic heating of grape juice. *Journal of Food Engineering, 85*, 410–417.

Jequier, E., & Constant, F., (2010). Water as an essential nutrient: The physiological basis of hydration. *European Journal of Clinical Nutrition, 64*, 115–123.

Lee, A., Elam, J. W., & Darling, S. B., (2016). Membrane materials for water purification: Design, development, and application. *Environmental Science Water Research and Technology, 2*, 17–42.

Lee, C. H., & Yoon, S. W., (1999). Effect of ohmic heating on the structure and permeability of the cell membrane of *Saccharomyces cerevisiae*. *IFT Annual Meeting*. Chicago.

Luis, F., Machado, R. N., Pereira, R. C., Martins, J. A., & Teixeira, A. A. V., (2009). Moderate electric fields can inactivate *Escherichia coli* at room temperature. *IBB – Institute for Biotechnology and Bioengineering, Centre of Biological Engineering*. Universidade do Minho, 4710-057 Braga, Portugal.

Machado, L. F., Pereira, R. N., Martins, R. C., Teixeira, J. A., & Vicente, A. A., (2010). Moderate electric fields can inactivate *Escherichia coli* at room temperature. *Journal of Food Engineering, 96*, 520–527.

Mohsin, G. F., (2011). Ohmic milk pasteurizer design, manufacture and studying its pasteurization efficiency. *M.Sc. Thesis* (p. 109). Food Science, Agriculture College, Basrah University.

Pereira, R., Pereira, M., Teixeira, J., & Vicente, A., (2007). Comparison of chemical properties of food products processed by conventional and ohmic heating. *Chemical Papers, 61*(1), 30–35.

Phillips, P. A., Rolls, B. J., Ledingham, J. G., et al., (1984). Reduced thirst after water deprivation in healthy elderly men. *The New England Journal of Medicine, 311*(12), 753–759.

Popkin, B. M., D'Anci, K. E., & Rosenberg, I. H., (2010). Water, hydration and health. *Nutrition Reviews, 68*(8), 439–458.

Rastogi, R. P., Richa, K. A., Tyagi, M. B., & Sinha, R. P., (2010). Molecular mechanisms of ultraviolet radiation-induced DNA damage and repair. *Journal of Nucleic Acids*, 1–32. http://dx.doi.org/10.4061/2010/592980 (Accessed on 8 November 2019).

SPSS (Statistical Package for the Social Science), (2001). *SPSS Statistical Package for Windows*. Ver. 11.0 Chicago: SPSS, Inc.

Sussman, S., (1978). Use of chlorine dioxide in water and wastewater treatment. In: Rice, R. G., & Cotruvo, J. A., (eds.), *Ozone/Chlorine Dioxide Oxidation Products of Organic Materials* (pp. 344–355). Proceedings of a Conference held in Cincinnati, Ohio – 1976. Sponsored by the International Ozone Institute, Inc., and the U.S. Environmental Protection Agency. Ozone Press International, Cleveland, Ohio.

Symons, J. M., Carswell, J. K., Clark, R. M., Dorsey, P., Geldreich, E. E., Heffernam, W. P., Hoff, J. C., Love, O. T., McCabe, L. J., & Stevens, A. A., (1977). Ozone, chlorine dioxide and chloramines as alternatives to chlorine for disinfection of drinking water: State of the art. *Water Supply Research Division* (p. 84). U.S. Environmental Protection Agency, Cincinnati, Ohio.

Trakhtman, N. N., (1946). Chlorine dioxide in water disinfection. *Gigiena Sanitariia, 11*(10), 10–13.

USCB (United States Census Bureau), (2008). *Current Housing Reports, Series H150/07.* American Housing Survey for the United States: 2007. U.S. Government Printing Office. Washington, DC, USA. URL: http://www.census.gov/prod/2008pubs/h150–07. pdf (Accessed on 8 November 2019).

Wang, W. C., & Sastry, S. K., (1993). Salt diffusion into vegetable tissue as a pre-treatment for Ohmic heating: Determination of parameters and mathematical model verification. *Journal of Food Engineering, 20,* 311–323.

Wimalawansa, S. J., (2013). Purification of contaminated water with reverse osmosis: Effective solution of providing clean water for human needs in developing countries. *International Journal of Emerging Technology and Advanced Engineering, 3*(12), 75–89.

Wouters, P. C., Alvarez, I., & Raso, J., (2001). Critical factors determining inactivation kinetics by pulsed electric field food processing. *Trends in Food Science and Technology, 12,* 112–121.

# CHAPTER 5

# Removal Cholesterol from Minced Meat Using Supercritical $CO_2$

ASAAD REHMAN SAEED AL-HILPHY, MUNIR ABOOD JASIM AL-TAI, and HASSAN HADI MEHDI AL RUBAIY

*Department of Food Science, College of Agriculture, University of Basrah, Basrah City, Iraq, E-mail: aalhilphy@yahoo.co.uk (A.R.S. Al-Hilphy); E-mail: dr.munir2000@yahoo.com (M.A.J. Al-Tai); E-mail: drhassanhadi78@gmail.com (H.H.M. AlRubaiy)*

## 5.1 INTRODUCTION

The cholesterol is one of the steroid compounds which have a high molecular weight. Cholesterol color is white, and it doesn't have taste and odor, as well as its constitution is fatty. Cholesterol can be classified as one of the fat types that don't dissolve in water; also, it's an alcoholic compound because it contains the hydroxyl group (Bansal et al., 2005).

All parts of the human body having cholesterol and all body cells can be making it, as well as the lever is the essential source for making cholesterol in the human body. Cholesterol is important and essential for building the body because it enters into the construction of tissues and cells. Blood serum contains 0.16–0.24 g/100 ml (Dora et al., 2003). The human body can be making 1–2 g of cholesterol daily and removes 0.1–0.8 g/day via skin and faces, while the body needs 0.2–0.3 g. of cholesterol for healthy persons, but the patients need less than 0.2 g/day of cholesterol. On the other hand, the quantity of cholesterol daily needed for men is less than women.

Previous studies indicate that scientists have interested with how reducing cholesterol levels in human blood and food. Some of the researchers have found that possibility using types of Bactria as an alternate from medicines for reducing cholesterol levels in the blood such as

*Lactobacillus* (Yu et al., 1994). There are four methods for the reduction of cholesterol in food, such as physical, chemical, biological, and enzymatic methods (Bradley et al., 1989). These methods have many problems, such as removes odor and natural compounds of food. Its cost is highly expensive and causes toxic effects in the food, especially at using different chemical solvents (Neves, 1996).

Nowadays, supercritical fluids have used as novel methods for removing or reducing cholesterol in the food (Cully et al., 1989; Saldan et al., 1997). Supercritical fluids have used for removing cholesterol from dried meat, chicken, ham meat, fish, and eggs (Froning et al., 1992, 1994). In addition, the production of fatty milk without cholesterol (reduced to 90%) (Bradley, 1991). Also, supercritical fluids have been used for fractionation of butter (Shishikura et al., 1996).

## 5.2   SUPERCRITICAL FLUIDS

Supercritical fluids are fluids which have temperature and pressure higher than its critical temperature and pressure (Verma et al., 2018). For example, the critical temperature of $CO_2$ is 31.1°C, and the critical pressure is 73.76 bar (Hawthorne, 1990; McKinnon and Parratt, 2002). The character of solubility for many fluids at supercritical conditions is a general character for it. When pressure increases with the increase of temperature, the solubility has significantly increased. On the other hand, the solubility decreases with the decrease in density (McKinnon and Parratt, 2002). The characters of Viscosity and diffusivity of supercritical $CO_2$ (SC-$CO_2$) are close to gases, but the density is close to liquids.

Bravi et al. (2007) have demonstrated that the extraction by using supercritical fluids is faster because of the reduction of its viscosity and increasing its diffusivity. Besides, its physiochemical such as density, diffusivity, dielectric constant, and viscosity can be controlled easily via changing pressure or temperature without change supercritical fluid state. Dixon and Johnston (1997) and Brunner (2005) have stated that the density of the supercritical fluid is the same as liquids, while their viscosity and diffusivity are close to gases. As a result, it can be spread faster in the solid materials compare with liquids. It can be seen from Figure 5.1, which illustrates the changing phase of $CO_2$. There are three phases: solid, gas, and liquid, as well as the supercritical region. The points presented between two phases are known as an equilibrium case (Akgerman et al., 1991). There is only one phase that lays in the critical region is a known supercritical fluid (no gas and no liquid). In this region, the supercritical fluids do not condensate and boiling whatever temperature increases (Hawthorne, 1990).

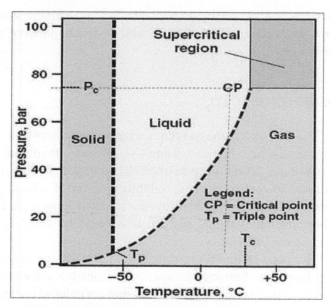

**FIGURE 5.1** Supercritical CO$_2$ diagram (Bravi et al., 2007).

A critical temperature is a higher temperature which enables the gas to convert to liquid by compression with increasing pressure. Critical pressure is a higher pressure to evaporate fluids as a gas with increasing fluids temperature (Hawthorne, 1990).

Burnner (2005) has stated that the solubility power of SC-CO$_2$ can be summarized in the following:

1. Solubility of polar and non-polar compounds is little.
2. Solubility power of reduced molecular weight compounds is high, and reduced with the increase of molecular weight.
3. SC-CO$_2$ has a high ability in the solubility with middle molecular weight oxygenates organic compounds.
4. Solubility of free fatty acids and glycerides is little.

5. Paints solubility is little.
6. Water has reduced solubility power less than 0.5 w/w at temperature of 100°C.
7. Proteins, polysaccharides and mineral salts are not soluble in SC-$CO_2$.

SC-$CO_2$ has a high ability to separate the non-volatile compounds which have higher molecular weight and/or much polarity with the increase pressure.

### 5.2.1   BENEFITS OF SC-CO$_2$

It's known that $CO_2$ gas (in the gas state) is nonactive as a solvent for fluids and solid materials, but when increasing $CO_2$ pressure, its solubility power increases (Gabriela, 2004). In spite of SC-$CO_2$ properties rarely converge with traditional solvent properties. Solubility power of a large group of organic compounds was evaluated by the scientist Francis in 1954 that compared between liquids $CO_2$ and 261 of other materials (Akgerman et al., 1991; Gabriela, 2004).

SC-$CO_2$ is a good solvent for organic compounds. Besides, its non-toxic, chemically inert, non-burning, inexpensive, available commercial with high purity, its critical temperature and pressure are reduced compared with other fluids. Its characteristics are converged for organic solvent. It can be removing from product easily. It's a non-oxidizer media and safety media.

### 5.2.2   DRAWBACKS OF SC-CO$_2$

SC-$CO_2$ has not enough dissolution power for polar compounds at a pressure of 30–80 bar (Foster et al., 1993). On the other hand, polar solvent (such as ethanol) can be added to the enhancement of solubility (Gabriela, 2004).

### 5.2.3   PHYSICAL PROPERTIES OF SC-CO$_2$

Supercritical fluids have Physical properties lay between liquids and gases. It can be spread on the surface easily compared with other real fluids, because the surface tension of Supercritical fluids is very little compared with the other real fluids (Gabriela, 2004). It can be seen from Table 5.1 that viscosity of SC-$CO_2$ is like the viscosity of $CO_2$ gas, but it less than the liquid. SC-$CO_2$ diffusivity lies between liquid and gas (Taylor, 1996). It can be control of Supercritical fluids solvents via changing pressure and temperature, also the solubility power of Supercritical fluids direct proportion with density (Johnston, 1989). Gabriela (2004) has announced that the solubility power of Supercritical fluids increases with the increase of density, also the density increases with the increase of pressure.

**TABLE 5.1**  Physical Properties of $CO_2$ States

| $CO_2$ | Density (g/cm³) | Kinematic Viscosity (g/cm.sec) | Diffusivity (cm²/sec) |
|---|---|---|---|
| Gas | 0.0006–0.002 | 0.0001–0.0030 | 0.100000–0.4 |
| Supercritical | 0.2000–0.500 | 0.0001–0.003 | 0.0007 |
| Liquid | 0.6000–1.600 | 0.0020–0.030 | 0.000002–0.00002 |

(*Source*: Taylor, (1996)).

Density and viscosity of SC-$CO_2$ are function of pressure at different temperatures ranged between 40–100°C. This means when the inserted pressure increases from 7–62 Mpa, the density and viscosity increase. Kumovo and Hassan (2007) have mentioned that density and viscosity of SC-$CO_2$ increase with increase of inserted pressure, and reduce with increase of temperature. When pressure of 7.5, 12.5 and 20 Mpa, the density reached 0.6618, 0.8170 and 0.8911 g/cm³ respectively, and the viscosity reached $5.17 \times 10^2$, $7.7 \times 10^2$ and $9.04 \times 10^2$ g/cm.s, respectively. When temperature of 30, 45 and 60°C, density reached 0.6618, 0.2093 and 0.1728 g/cm³, respectively, and the viscosity reached $5.17 \times 10^{-2}$, $2.09 \times 10^{-2}$ and $1.73 \times 10^{-2}$ g/cm.s, respectively. Also, they stated that the diffusion coefficient of SC-$CO_2$ reduces with the increases of pressure and temperature.

Dynamic viscosity can be calculated from the following equation (Ouyang, 2011):

$$\mu = C_0 + C_1 P + C_2 P^2 + C_3 P^3 + C_4 P^4 \qquad (1)$$

where, μ: dynamic viscosity, P: pressure (Pa), and c: constant.

C is calculated as follow:

$$C_i = d_{i0} + d_{i1}T + d_{i2}T^2 + d_{i3}T^3 + d_{i4}T^4 \tag{2}$$

where, T: temperature (°C), I: 1, 2, 3,...

The constants related with Equation (2) are showed in Table 5.2.

**TABLE 5.2**  Constants Related With Equation (2)

| $d_{i4}$ | $d_{i3}$ | $d_{i2}$ | $d_{i1}$ | $d_{i0}$ | |
|---|---|---|---|---|---|
| −1.824 E−09 | 8.331 E−07 | −1.004 E−04 | 3.083 E−03 | 1.856 E−02 | i = 0 |
| 1.463 E−12 | −6.141 E−10 | 7.524 E−08 | −3.174 E−06 | 6.519 E−05 | i = 1 |
| −3.852 E−16 | 1.530 E−13 | −1.830 E−11 | 7.702 E−10 | −1.310 E−08 | i = 2 |
| 4.257 E−20 | −1.632 E−17 | 1.921 E−15 | −8.113 E−15 | 1.335 E−12 | i = 3 |
| −1.691 E−24 | 6.333 E−22 | −7.370 E−20 | 3.115 E−18 | −5.047 E−17 | i = 4 |

(*Source*: Adapted from Ouyang, 2011.)

Density can be calculated from the following equation (Ouyang, 2011):

$$\rho = A_0 + A_1P + A_2P^2 + A_3P^3 + A_4P^4 \tag{3}$$

where, $\rho$: density (kg/m³), $A_1$, $A_2$, $A_3$, $A_4$ are constants can be calculated as follow:

$$A_i = b_{i0} + b_{i1}T + b_{i2}T^2 + b_{i3}T^3 + b_{i4}T^4 \tag{4}$$

The constants related with Equation (4) are shown in Table 5.3.

**TABLE 5.3**  Constants Related With Equation (4)

| $b_{i4}$ | $b_{i3}$ | $b_{i2}$ | $b_{i1}$ | $b_{i0}$ | |
|---|---|---|---|---|---|
| 3.439 E−05 | −4.651 E−03 | −2.254 E−02 | 2.730 E + 00 | 6.897 E + 02 | i = 0 |
| −1.888 E−08 | 2.274 E−06 | 5.982 E−05 | −6.547 E−03 | 2.213 E−01 | i = 1 |
| 3.893 E−12 | −4.079 E−10 | −2.311 E−08 | 2.019 E−06 | −5.118 E−05 | i = 2 |
| −3.560 E−16 | 3.171 E−14 | 3.121 E−12 | −2.415 E−10 | 5.517 E−09 | i = 3 |
| 1.215 E−20 | −8.957 E−19 | −1.406 E−16 | 1.010 E−14 | −2.184 E−13 | i = 4 |

(*Source*: Adapted from Ouyang, 2011.)

Figure 5.2 illustrates that the density of SC-$CO_2$ significantly increases as the pressure increases and significantly decreases as the temperature rises at all temperatures (35, 45, and 55°C) (Al-Hilphy et al., 2016). Figure 5.3 shows that the viscosity of SC-$CO_2$ increases with the significantly increase of the pressure at all temperatures (35, 45, and 55°C). This because of the increasing SC-$CO_2$ density as pressure increases. On the other hand, the viscosity significantly decreases with the increase of temperature.

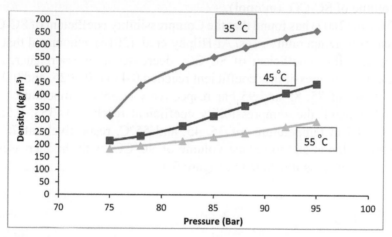

**FIGURE 5.2** Density of SC-$CO_2$ vs. pressure at different temperatures (*Source*: Reprinted from Al-Hilphy et al., 2017).

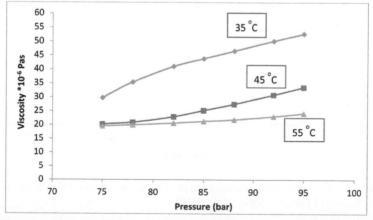

**FIGURE 5.3** Viscosity of SC-$CO_2$ vs. pressure at different temperatures (*Source*: Reprinted from Al-Hilphy et al., 2017).

### 5.2.4  COMPRESSIBILITY COEFFICIENT OF SC-CO₂

Compressibility coefficient of SC-CO$_2$ is calculated from the following equation (Marini, 2007):

$$Z = \frac{PV}{RT} \tag{5}$$

where, R: gas constant (J/mol.K), T: temperature (K), P: pressure (N/m²), V: volume of SC-CO$_2$ (m³/mol).

Marini (2007) has found that the Compressibility coefficient of SC-CO$_2$ increases as temperature rises. Al-Hilphy et al. (2016) announced that the Compressibility coefficient of SC-CO$_2$ decreases as pressure increases, where the Compressibility coefficient reached 0.4354, 0.3025, and 0.0254 at pressure of 75, 85 and 95 bar respectively at 35°C temperatures. On the other hand, the compressibility coefficient reached 0.4353, 0.5871, and 0.6607 at temperatures of 35, 45, and 55°C, respectively at 75 bar pressure. This is due to reduce volume of SC-CO$_2$ with the increases of pressure highly, as illustrated in Figure 5.4.

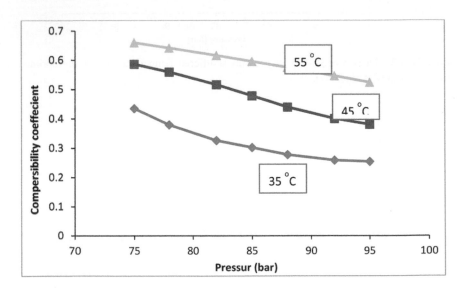

**FIGURE 5.4**   Compressibility coefficient of SC-CO$_2$ vs. pressure at different temperatures (*Source*: Reprinted from Al-Hilphy et al., 2017).

## 5.3 REMOVAL CHOLESTEROL USING SC-$CO_2$

SC-$CO_2$ is used for removing cholesterol from foods because it has a high ability to soluble fats and cholesterol (Kosal et al., 1992). Yun et al. (1991) have used a continuous SC-$CO_2$ system to remove the cholesterol and triglycerides. Also, Neves (1996) has manufactured a continuous apparatus for extracting cholesterol by using SC-$CO_2$. Chrastil (1982) has manufactured a batch device for extracting cholesterol by using SC-$CO_2$ technique.

Al-Hilphy et al. (2016) have designed an apparatus used for removing cholesterol from minced meatworks by the static and dynamic methods. It consists of a booster pump that operates by compressed air (10 bar). Booster pump rises the pressure of $CO_2$ up to 200 bar, air compressor, $CO_2$ bottle, control valves (maximum pressure is 240 bar), safety valve, heating part is used for heating removal cholesterol cylinder, cooling part is used for cooling $CO_2$ before entrance to the booster pump, Adsorption unit of Cholesterol by $CaCO_3$ and temperature and pressure gauges, as shown in Figure 5.5. The operation chart of batch and dynamic SC-$CO_2$ is illustrated in Figure 5.6.

**FIGURE 5.5** Supercritical $CO_2$ apparatus (*Source*: Reprinted from Al-Hilphy et al., 2017).

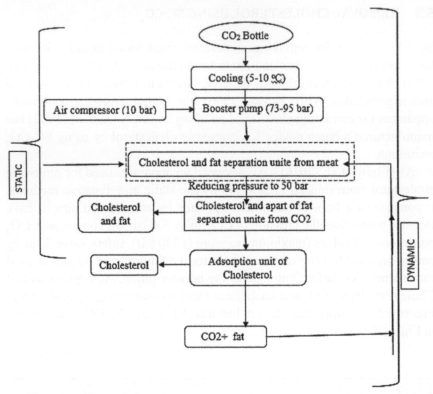

**FIGURE 5.6**  Operation chart of batch and dynamic SC-CO$_2$ (*Source*: Reprinted from Al-Hilphy et al., 2017).

SC-CO$_2$ technique can be used for removing fats and cholesterol from minced meats, where fat ratio of beef meat was 5.77% and reached to 1.1, 0.98 and 9.42% after using SC-CO$_2$ at pressure of 75, 85, 95 bar, respectively at temperature of 35°C, also, The relationship between removing fats and temperature is reversal (Al-Hilphy et al., 2017). Kumovo and Hassan (2007) reported that the increase of temperature leads to a huge reduce in the density and viscosity, which causes molecules of CO$_2$ is separated and prevents dissolve of fats in SC-CO$_2$. On the other hand, when holding time increases, removing fat and cholesterol rise. The static method has a higher removal cholesterol percentage compare with the dynamic method. Tables 5.4 and 5.5 show the removal of cholesterol percentage using SC-CO$_2$ with static and dynamic methods.

**TABLE 5.4** Removal Cholesterol Percentage (%) from Minced Beef Meat Using Static Method SC-$CO_2$

| Pressure Bar | 35°C | | | | 45°C | | | | 55°C | | | |
|---|---|---|---|---|---|---|---|---|---|---|---|---|
| | Holding Time (min) | | | | Holding Time (min) | | | | Holding Time (min) | | | |
| | 20 | 40 | 60 | 80 | 20 | 40 | 60 | 80 | 20 | 40 | 60 | 80 |
| 75 | 15.30 | 16.35 | 61.28 | 79.68 | 14.50 | 24.37 | 47.24 | 62.44 | 13.49 | 26.48 | 43.58 | 56.83 |
| 85 | 16.39 | 34.09 | 62.25 | 82.32 | 15.30 | 25.88 | 51.42 | 63.57 | 13.61 | 26.93 | 46.78 | 58.86 |
| 95 | 17.45 | 39.44 | 67.37 | 87.10 | 16.81 | 37.32 | 52.21 | 67.57 | 14.42 | 32.28 | 51.48 | 64.36 |

(*Source*: Reprinted from Al-Hilphy et al., (2017)).

**TABLE 5.5** Removal Cholesterol Percentage (%) from Minced Beef Meat Using Dynamic Method SC-$CO_2$

| Pressure Bar | 35°C | | | | 45°C | | | | 55°C | | | |
|---|---|---|---|---|---|---|---|---|---|---|---|---|
| | Holding Time (min) | | | | Holding Time (min) | | | | Holding Time (min) | | | |
| | 20 | 40 | 60 | 80 | 20 | 40 | 60 | 80 | 20 | 40 | 60 | 80 |
| 75 | 14.84 | 38.27 | 41.66 | 51.00 | 6.27 | 15.38 | 31.26 | 43.99 | 5.85 | 13.72 | 28.41 | 42.02 |
| 85 | 15.78 | 39.00 | 45.51 | 56.19 | 11.03 | 20.33 | 36.62 | 50.42 | 14.86 | 25.45 | 44.05 | 49.57 |
| 95 | 17.01 | 36.23 | 59.64 | 70.69 | 14.42 | 28.41 | 44.76 | 60.85 | 10.99 | 25.97 | 35.59 | 52.15 |

(*Source*: Reprinted from Al-Hilphy et al., (2017)).

## 5.4   CHOLESTEROL SOLUBILITY IN SC-CO$_2$

Chrastil (1982) has studied the cholesterol of solubility in SC-CO$_2$ using pressure ranged between 100–250 bar at a temperature of 20, 40, and 80°C, and noticed that increasing cholesterol solubility with pressure at static temperature. This because increasing CO$_2$ density with pressure, as a result, increases solvent power. When the temperature increases, the cholesterol solubility decreases.

Huang et al. (2004) have cholesterol solubility increases with the increase of the SC-CO$_2$ pressure. Moreover, they approved that the temperature has a negative effect on the cholesterol solubility in SC-CO$_2$, because increasing temperature leads to increase the vapor pressure of cholesterol, as a result cholesterol solubility increases, but reducing density of SC-CO$_2$ leads to decrease the solubility. Table 5.6 illustrates the empirical equations for solute solubility in SC-CO$_2$.

**TABLE 5.6**   Empirical Equations for Solute Solubility in SC-CO$_2$

| Equations | References |
|---|---|
| $\ln S = k \ln \rho + A + \dfrac{P}{T}$ | Chrastil, 1982 |
| $S = A - BT + CT + DP$  (general equation) | Chrastil, 1982 |
| $\ln y_1 = A + B\rho + \dfrac{C}{T}$ | Kumar and Johnston, 1988 |
| $\ln S = A + \dfrac{B}{T} + C \ln \rho + \dfrac{D}{T^2}$ | Del Valle and Aguilera, 1988 |
| $T \ln(y_2 P) = A + B\rho + CT$ | Méndez-Santiago and Teja, 2000 |
| $\ln\left(\dfrac{y_2 P}{P_{ref}}\right) = A + \dfrac{P}{T} + \left(\rho - \rho_{ref}\right)$ | Bartle et al., 1991 |
| $y_2 = A + BP + CP^2 + DPT\left(1 - y_2\right) + ET + FT^2$ | Yun et al., 1991 |
| $\ln y_2 = A + \dfrac{B}{T} + \left(C + \dfrac{D}{T}\right)\ln \rho$ | Taylor, 1996 |

**TABLE 5.6** (Continued)

| Equations | References |
|---|---|
| $\ln y_2 = A + BT + CT^2 + DPT + EP + BT^2$ | Gordillo et al., 1999 |
| $y_2 = A + BP + CP^{2+} DPT + \dfrac{ET}{P} + F \ln \rho$ | Jouyban et al., 2002 |
| $\ln \rho = -27.091 + 0.609\sqrt{T} + \dfrac{3966.170}{T} - \dfrac{3.445P}{T} + 0.402\sqrt{P}$ | Jouyban et al., 2002 |
| $\ln y_2 = (A + B\rho)\ln \rho + \dfrac{C}{T} + D$ | Adachi and Lu, 1983 |
| $\ln y_2 = A\ln\left(\rho T + \dfrac{P}{T} + C\right)$ | Garlapati and Madras, 2009 |
| $\ln y_2 = A + (C\rho)\ln \rho + \dfrac{D}{T} + E\ln(\rho T)$ | Garlapati and Madras, 2010 |
| $\ln y_2 = A + BP^2 + CT^2 + D\ln \rho$ | Jafari Nejad et al., 2010 |
| $\ln y_2 = \dfrac{A}{T} + BP + \dfrac{CP^2}{T} + (D + EP)\ln \rho$ | Khansary et al., 2015 |
| S=−0.00929−0.00152T+0.002046t+0.000816P(*static method*) (for cholesterol) (a) | Al-Hilphy et al., 2016 |
| S=−0.0366−0.00118T+0.001532t+0.001049P (*dynamic method*) (for cholesterol)(b) | Al-Hilphy et al., 2016 |

A, B. C. D, E, F, k are constants. S, y$_2$ T, P, $\rho$, t are solubility (g/l), solute solubility (mole fraction), temperature, pressure, density, and holding time, respectively.

Holding time of solute under SC-CO$_2$ pressure has a high effect on the cholesterol solubility, as illustrated in Tables 5.7 and 5.8. Cholesterol solubility was increased with the increase of the holding time. Also, cholesterol solubility increases at using the static method of SC-CO$_2$ compare with the dynamic method because the mass transfer coefficient by using the static method is higher than the dynamic method. Also, the mass transfer coefficient reduced with increasing pressure in the case of a static method (Al-Hilphy et al., 2016).

**TABLE 5.7** The Cholesterol Solubility at Different Temperatures, Pressures, and Holding Times us Static Method Using Different Equations

| Temperature (°C) | Time (min.) | Pressure Bar | Solubility Using Equation of Al-Hilphy et al. (2016) (g/l) | Solubility Using Equation of Chrastil (1982) (g/l) | Experimental Solubility (g/l) | Mass Transfer Coefficient (km) m²/s |
|---|---|---|---|---|---|---|
| 35 | 20 | 75 | 0.039616709 | 0.033545342 | 0.03354534 | 3.77567 E–05 |
| 35 | 20 | 85 | 0.047773457 | 0.035891055 | 0.035925693 | 1.85394 E–05 |
| 35 | 20 | 95 | 0.055930206 | 0.038139817 | 0.038129723 | 1.34681 E–05 |
| 35 | 40 | 75 | 0.080541142 | 0.070354019 | 0.072071788 | 3.77567 E–05 |
| 35 | 40 | 85 | 0.08869789 | 0.07841199 | 0.074716625 | 1.85394 E–05 |
| 35 | 40 | 95 | 0.096854639 | 0.086441968 | 0.086442065 | 1.34681 E–05 |
| 35 | 60 | 75 | 0.121465575 | 0.134313452 | 0.134313602 | 3.77567 E–05 |
| 35 | 60 | 85 | 0.129622324 | 0.161596584 | 0.137178841 | 1.85394 E–05 |
| 35 | 60 | 95 | 0.137779072 | 0.190827902 | 0.190891058 | 1.34681 E–05 |
| 35 | 80 | 75 | 0.162390008 | 0.174647304 | 0.174647355 | 3.77567 E–05 |
| 35 | 80 | 85 | 0.170546757 | 0.183020373 | 0.180421914 | 1.85394 E–05 |
| 35 | 80 | 95 | 0.178703505 | 0.190890833 | 0.190891058 | 1.34681 E–05 |
| 45 | 20 | 75 | 0.024418084 | 0.031782088 | 0.031782116 | 3.64802 E–05 |
| 45 | 20 | 85 | 0.032574832 | 0.034355559 | 0.03354534 | 3.18245 E–05 |
| 45 | 20 | 95 | 0.040731581 | 0.036846815 | 0.036851385 | 2.67611 E–05 |
| 45 | 40 | 75 | 0.065342517 | 0.052584629 | 0.053425693 | 3.64802 E–05 |
| 45 | 40 | 85 | 0.073499265 | 0.066365311 | 0.056731738 | 3.18245 E–05 |
| 45 | 40 | 95 | 0.081656014 | 0.081813502 | 0.081813602 | 2.67611 E–05 |

**TABLE 5.7** (Continued)

| Temperature (°C) | Time (min.) | Pressure Bar | Solubility Using Equation of Al-Hilphy et al. (2016) (g/l) | Solubility Using Equation of Chrastil (1982) (g/l) | Experimental Solubility (g/l) | Mass Transfer Coefficient (km) m²/s |
|---|---|---|---|---|---|---|
| 45 | 60 | 75 | 0.10626695 | 0.10354534 | 0.10354534 | 3.64802 E-05 |
| 45 | 60 | 85 | 0.114423699 | 0.109072698 | 0.112714106 | 3.18245 E-05 |
| 45 | 60 | 95 | 0.122580447 | 0.114293843 | 0.114301008 | 2.67611 E-05 |
| 45 | 80 | 75 | 0.147191383 | 0.136870272 | 0.136870277 | 3.64802 E-05 |
| 45 | 80 | 85 | 0.155348132 | 0.142729181 | 0.139338791 | 3.18245 E-05 |
| 45 | 80 | 95 | 0.16350488 | 0.148210784 | 0.148110831 | 2.67611 E-05 |
| 55 | 20 | 75 | 0.009219458 | 0.029573139 | 0.029578086 | 3.34552 E-05 |
| 55 | 20 | 85 | 0.017376207 | 0.030626595 | 0.029842569 | 2.5371 E-05 |
| 55 | 20 | 95 | 0.025532955 | 0.031605765 | 0.031605793 | 1.7986 E-05 |
| 55 | 40 | 75 | 0.050143892 | 0.058036016 | 0.058054156 | 3.64802 E-05 |
| 55 | 40 | 85 | 0.05830064 | 0.064416191 | 0.059023929 | 3.18245 E-05 |
| 55 | 40 | 95 | 0.066457389 | 0.070749326 | 0.07074937 | 2.67611 E-05 |
| 55 | 60 | 75 | 0.091068325 | 0.095522578 | 0.09552267 | 3.64802 E-05 |
| 55 | 60 | 85 | 0.099225073 | 0.102571498 | 0.102531486 | 3.18245 E-05 |
| 55 | 60 | 95 | 0.107381822 | 0.109352243 | 0.112846348 | 2.67611 E-05 |
| 55 | 80 | 75 | 0.131992758 | 0.124560198 | 0.124571788 | 3.34552 E-05 |
| 55 | 80 | 85 | 0.140149507 | 0.132991314 | 0.129023929 | 2.53710 E-05 |
| 55 | 80 | 95 | 0.148306255 | 0.141057919 | 0.141057935 | 1.79860 E-05 |

**TABLE 5.8** The Cholesterol Solubility at Different Temperatures, Pressures, and Holding Times Using Dynamic Method Using Different Equations

| Temperature (°C) | Time (min.) | Pressure Bar | Solubility Using Equation of Al-Hilphy et al. (2016) (g/l) | Solubility Using Equation of Chrastil (1982) (g/l) | Experimental Solubility (g/l) | Mass Transfer Coefficient (km) m²/s |
|---|---|---|---|---|---|---|
| 35 | 20 | 75 | 0.031561835 | 0.033986145 | 0.033986146 | 1.75066 E−05 |
| 35 | 20 | 85 | 0.042053019 | 0.036159144 | 0.036146096 | 9.55331 E−06 |
| 35 | 20 | 95 | 0.052544203 | 0.038231269 | 0.038967254 | 6.91773 E−06 |
| 35 | 40 | 75 | 0.062197366 | 0.074407825 | 0.07440806 | 1.75066 E−05 |
| 35 | 40 | 85 | 0.07268855 | 0.078794813 | 0.067267003 | 9.55331 E−06 |
| 35 | 40 | 95 | 0.083179733 | 0.08295958 | 0.082959698 | 6.91773 E−06 |
| 35 | 60 | 75 | 0.092832896 | 0.096448357 | 0.096448363 | 1.75066 E−05 |
| 35 | 60 | 85 | 0.10332408 | 0.115238001 | 0.104206549 | 9.55331 E−06 |
| 35 | 60 | 95 | 0.113815264 | 0.135237615 | 0.136561713 | 6.91773 E−06 |
| 35 | 80 | 75 | 0.123468426 | 0.116769385 | 0.116769521 | 1.75066 E−05 |
| 35 | 80 | 85 | 0.13395961 | 0.138664403 | 0.128671285 | 9.55331 E−06 |
| 35 | 80 | 95 | 0.144450794 | 0.161834367 | 0.16186398 | 6.91773 E−06 |
| 45 | 20 | 75 | 0.019801497 | 0.025698813 | 0.025698992 | 2.3369 E−05 |
| 45 | 20 | 85 | 0.030292681 | 0.02932332 | 0.025258186 | 1.33846 E−05 |

**TABLE 5.8** (Continued)

| Temperature (°C) | Time (min.) | Pressure Bar | Solubility Using Equation of Al-Hliphy et al. (2016) (g/l) | Solubility Using Equation of Chrastil (1982) (g/l) | Experimental Solubility (g/l) | Mass Transfer Coefficient (km) $m^2/s$ |
|---|---|---|---|---|---|---|
| 45 | 20 | 95 | 0.040783865 | 0.033016577 | 0.033016373 | 9.41969 E–06 |
| 45 | 40 | 75 | 0.050437028 | 0.059464718 | 0.059464736 | 2.3369 E–05 |
| 45 | 40 | 85 | 0.060928212 | 0.062346013 | 0.04654918 | 1.33846 E–05 |
| 45 | 40 | 95 | 0.071419395 | 0.065055606 | 0.065062972 | 9.41969 E–06 |
| 45 | 60 | 75 | 0.081072558 | 0.081790824 | 0.071586902 | 2.3369 E–05 |
| 45 | 60 | 85 | 0.091563742 | 0.085957179 | 0.085957179 | 1.33846 E–05 |
| 45 | 60 | 95 | 0.102054926 | 0.089884065 | 0.102487406 | 9.41969 E–06 |
| 45 | 80 | 75 | 0.111708088 | 0.09893401 | 0.100724181 | 2.3369 E–05 |
| 45 | 80 | 85 | 0.122199272 | 0.118483554 | 0.115447103 | 1.33846 E–05 |
| 45 | 80 | 95 | 0.132690456 | 0.139337937 | 0.139338791 | 9.41969 E–06 |
| 55 | 20 | 75 | 0.008041159 | 0.013400479 | 0.013400504 | 3.64802 E–05 |
| 55 | 20 | 85 | 0.018532343 | 0.018686052 | 0.034030227 | 3.18245 E–05 |
| 55 | 20 | 95 | 0.029023527 | 0.025196846 | 0.025170025 | 2.67611 E–05 |
| 55 | 40 | 75 | 0.03867669 | 0.040245599 | 0.040245592 | 3.64802 E–05 |
| 55 | 40 | 85 | 0.049167874 | 0.049460514 | 0.059023929 | 3.18245 E–05 |

**TABLE 5.8** *(Continued)*

| Temperature (°C) | Time (min.) | Pressure Bar | Solubility Using Equation of Al-Hilphy et al. (2016) (g/l) | Solubility Using Equation of Chrastil (1982) (g/l) | Experimental Solubility (g/l) | Mass Transfer Coefficient (km) m²/s |
|---|---|---|---|---|---|---|
| 55 | 40 | 95 | 0.059659058 | 0.059553977 | 0.059464736 | 2.67611 E–05 |
| 55 | 60 | 75 | 0.06931222 | 0.065116389 | 0.065062972 | 3.64802 E–05 |
| 55 | 60 | 85 | 0.079803404 | 0.073287038 | 0.08895466 | 3.18245 E–05 |
| 55 | 60 | 95 | 0.090294588 | 0.08150505 | 0.081505038 | 2.67611 E–05 |
| 55 | 80 | 75 | 0.09994775 | 0.096750139 | 0.09622796 | 3.64802 E–05 |
| 55 | 80 | 85 | 0.110438934 | 0.108089145 | 0.113507557 | 3.18245 E–05 |
| 55 | 80 | 95 | 0.120930118 | 0.119414358 | 0.119414358 | 2.67611 E–05 |

## 5.5  MATHEMATICAL MODELING

The solubility of cholesterol in SC-$CO_2$ is related with density of SC-$CO_2$. Chrastil (1982) has concluded the following equation for calculating solubility:

$$S = \rho^K \exp\left(\frac{A}{T} + B\right)$$

(6)

where, Sc: solubility ($kg/m^3$).

Mass transfer coefficient is calculated as follow (Norhuda and Mohd, 2009):

$$K_m = \frac{Sh\, D_{12}}{d_p}$$

(7)

where, $K_m$: Mass transfer coefficient, $Sh$: Sherwood number, $D_{12}$: diffusivity cholesterol coefficient ($m^2/s$), $d_p$: diameter of meat parts (m).

Sherwood number can be calculated from Equation (8) (Wakao and Kajuei, 1982):

$$Sh = 2 + 1.1 Re^{0.6} Sc^{0.3}$$

(8)

where, $Sh$: Sherwood number, $Re$: Reynold's number, and $S_c$: Schmidt number.

Reynold's number is calculated as below:

$$R_e = \frac{u d_p \rho}{\mu}$$

(9)

where, u: velocity of SC-$CO_2$ (m/sec), $d_p$: diameter of meat (m), $\rho$: density of SC-$CO_2$ ($kg/m^3$) and $\mu$: dynamic viscosity (Pa s).

Schmidt number is calculated from Equation (10) (Norhuda and Mohd, 2009):

$$S_c = \frac{\mu}{\rho D_{12}}$$

(10)

For calculating cholesterol diffusivity coefficient in SC-$CO_2$, the following equation is used (Catchpole and King, 1994):

$$D_{12} = 5.152 + D_c T_r \left( \rho_r^{-2\gamma} - 0.451 \right) \frac{K}{X} \tag{11}$$

where, $D_{12}$: cholesterol diffusivity coefficient, $D_c$: $CO_2$ diffusivity coefficient at critical point (m²/sec), $T_r$: reduced temperature of SC-$CO_2$ (°C), $\rho_r$: reduced pressure of SC-$CO_2$ (bar), K: correction factor, X:

$T_r$ and $\rho_r$ were calculated as bellow (Wang et al., 2015):

$$T_r = \frac{T}{T_C} \tag{12}$$

$$\rho_r = \frac{\rho}{\rho_c} \tag{13}$$

were, $T$: temperature of SC-$CO_2$ (°C), $T_c$: critical pressure of SC-$CO_2$ (bar), pressure SC-$CO_2$ (bar), $T_c$: critical pressure of SC-$CO_2$ (bar).

X is calculated from Equation (14) (Catchpole and King, 1994):

$$x = \frac{\left[ 2 + \left( V_{c2} / V_{c1} \right)^{1/3} \right]^2}{\left[ 1 + M_1 / M_2 \right]} \tag{14}$$

where, $V_{c1}$ molar volume of SC-$CO_2$ at critical point (cm³/mol), $V_{c2}$: Molar volume of cholesterol at critical point (cm³/mol), $M_1$: molecular weight of SC-$CO_2$ (g/mol), $M_2$: molecular weight of cholesterol (g/mol).

Correction factor can be calculated as bellow (Vedaraman et al., 2005):

$$K = 1 \pm 0.1 \ X < 2$$

$$K = X^{0.17} \pm 0.1 \ 2 < X < 10 \tag{15}$$

Diffusivity coefficient of SC-$CO_2$ is calculated from the following equation (Catchpole and King, 1994):

$$D_C = 4.30 \times 10^{-7} + M_1^{1/2} \frac{T_{C1}^{0.75}}{\sum V_1^{2/3} \rho_C} \tag{16}$$

where, $D_c$: Diffusivity coefficient of SC-$CO_2$ at critical point (m³/mol), $M_1$: molecular weight of SC-$CO_2$ at critical point $i$ (g/mol), $V_1$: molar volume of SC-$CO_2$ at critical point $i$ (m³/mol).

## 5.6  CHOLESTEROL DIFFUSIVITY IN SC-$CO_2$

It can be seen from Figure 5.7 that the cholesterol diffusivity coefficient increases as temperature rises, which indicates that temperature has a significant effect because increase of molecules movement and decreasing viscosity and density of $CO_2$. Moreover, the cholesterol diffusivity coefficient decreases with the increase pressure because decrease of molecules movement and increasing viscosity and density of $CO_2$ (Al-Hilphy et al., 2016). Vedaraman et al. (2005) found that the cholesterol diffusivity coefficient in SC-$CO_2$ reached $2.3×10^{-13}$ m²/s at all pressure values of 230 and 270 bar and temperature of 60°C, whereas, the cholesterol diffusivity coefficient increased to $2.8×10^{-13}$ m²/s at temperature of 70°C.

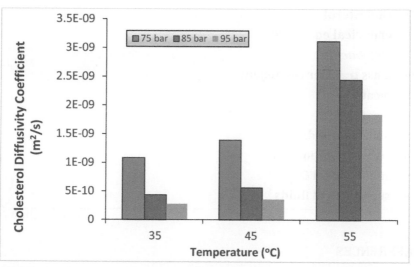

**FIGURE 5.7**   The cholesterol diffusivity coefficient in SC-$CO_2$ (*Source*: Reprinted from Al-Hilphy et al., 2016).

## 5.7 SUMMARY AND CONCLUSION

Supercritical dioxide carbon can be used for removing cholesterol from minced meat via changing temperature and pressure. Increasing the temperature of SC-CO$_2$ leads to a decrease in removal cholesterol, an increase of cholesterol diffusivity coefficient in SC-CO$_2$, but increasing pressure of SC-CO$_2$ leads to an increase of removal cholesterol and reduces the cholesterol diffusivity coefficient in SC-CO$_2$. Holding time has a significant effect on the removal of cholesterol. Cholesterol solubility increases at using the static method of SC-CO$_2$ compare with the dynamic method. The mass transfer coefficient is reduced with pressure. The Compressibility coefficient of SC-CO$_2$ increases as temperature rises and reduces with the increase of pressure. Density and viscosity of SC-CO$_2$ are the functions of pressure at different temperatures.

## KEYWORDS

- **butter**
- **cholesterol**
- **empirical equations**
- *lactobacillus*
- **mass transfer coefficient**
- **meat**
- **meat**
- **novel methods**
- **steroid compounds**
- **supercritical CO$_2$**
- **supercritical fluids**

## REFERENCES

Adachi, Y., & Lu, B. C. Y., (1983). Supercritical fluid extraction with carbon dioxide and ethylene. *Fluid Phase Equilibr., 14*, 147–156.

Akgerman, A., Roop, R. K., Hess, R. K., & Yeo, S. D., (1991). Supercritical extraction in environmental control. In: Bruno, J. T., & Bocca, R. J. F., (eds.), *Supercritical Fluid Technology: Reviews in Modern Theory and Applications* (pp. 479–509). CRC Press.

Al-Hilphy, A. R. S., Al-Tai, M. A. J., & AlRubaiy, H. H. M., (2016). A new empirical model for calculating solubility of cholesterol in supercritical dioxide carbon. *Journal of Biology, Agriculture and Healthcare, 6*(2), 1–11.

Al-Hilphy, A. R. S., Jassem, M. A., & Al-Rubaiy, H. H., (2017). Designing and manufacturing apparatus of removal cholesterol from meat using supercritical carbon dioxide. *Central Agency for Standardization and Quality Control*, No.4733 in 7–11–2016. Baghdad, Iraq.

Bansal, N., Cruicksank, J., MeElduff, P., & Durrington, P. N., (2005). Cord blood lipoproteins and prenatal influences. *Curr. Opin. Lipidol, 16*, 400–408.

Bartle, K. D., Clifford, A. A., Jafar, S. A., & Shilstone, G. F., (1991). Solubilities of solids and liquids of low volatility in supercritical carbon dioxide. *J. Phys. Chem. Ref. Data, 20*, 713–725.

Bradley, R. L., (1989). Removal of cholesterol from milk fat using supercritical carbon dioxide. *J. Dairy Sci., 72*, 2834–2840.

Bradley, R. L., (1991). Removing cholesterol from milk fat using supercritical carbon dioxide. In: Haberstroh, C., & Morris, C. E., (eds.), *Fat and Cholesterol Reduced Foods: Technologies and Strategies: Advances in Applied Biotechnology Series* (Vol. 12, pp. 221–223). TX: Gulf Pub., Woodlands.

Bravi, E., Perretti, G., Motanari, L., Favati, F., & Fantozzi, P., (2007). Supercritical fluid extraction for quality control in beer industry. *Journal of Supercritical Fluids, 42*, 342–346.

Brunner, G., (2005). Supercritical fluids: Technology and application to food processing. *J. Food Eng., 67*, 21–33.

Catchpole, O. J., & King, B., (1994). Measurement and correlation of binary diffusion coefficients in near critical fluids. *Ind. Eng. Chem. Res., 33*, 1828–1837.

Chrastil, J., (1982). Solubility of solids and liquids in supercritical gases. *J. Phys. Chem., 86*, 3016–3021.

Cully, J., Vollbrecht, R., & Schu, E., (1989). Process for Removing Cholesterol or Cholesterol Esters from Food. *German Patent Application DE 3929555 A1*.

Del Valle, J. M., & Aguilera, J. M., (1988). An improved equation for predicting the solubility of vegetable oils in supercritical CO$_2$. *Ind. Eng. Chem. Res., 27*, 1551–1559.

Dixon, D. J., & Johnston, K. P., (1997). Supercritical fluids. In: Ruthven, D. M., (ed.), *Encyclopedia of Separation Technology* (1st ed., pp. 1544–1569). John Wiley Interscience, New York.

Dora, L., Pereira, L., Mc Cartney, R., & Gibson, R., (2003). An *in vitro* study of the probiotic potential of abile salt. Hydrolyzire *Lactobacillus* fomenters stain and determination of its cholesterol lowering properties. *Appl. Environ. Microbiol., 69*(8), 4743–4752.

Foster, N. R., Singh, H., Yun, J., Tomasko, D. L., & Macnaughton, S. J., (1993). Polar and nonpolar co-solvent effects on the solubility of cholesterol in supercritical fluids. *Ind. Eng. Chem. Res., 32*(11), 2849–2853.

Froning, G. W., Cuppett, S. L., & Niemann, L., (1992). Extraction of cholesterol and other lipids from dehydrated beef using supercritical carbon dioxide. *J. Agric. Food Chem., 40*, 1204–1207.

Froning, G. W., Fieman, F., Wehiling, R. L., Cuppett, S., & Nielmann, L., (1994). Supercritical carbon dioxide extraction of lipids and cholesterol from dehydrated chicken meat. *Poultry Sci., 73*, 571–575.

Gabriela, I. B. S., (2004). Supercritical fluid technology: Computational and experimental equilibrium studies and design of supercritical extraction process. *PhD Dissertation.* University of Notre Dame, Indiana.

Garlapati, C., & Madras, G., (2009). Solubilities of solids in supercritical fluids using dimensionally consistent modified solvate complex models. *Fluid Phase Equilibr., 283,* 97–101.

Garlapati, C., & Madras, G., (2010). New empirical expressions to correlate solubilities of solids in supercritical carbon dioxide. *Thermochim. Acta, 500,* 123–127.

Gordillo, M. D., Blanco, M. A., Molero, A., & De la Ossa, E. M., (1999). Solubility of the antibiotic penicillin G in supercritical carbon dioxide. *J. Supercrit. Fluids, 15*(3), 183–190.

Hawthorne, S. B., (1990). Analytical scale supercritical fluid extraction, *Anal. Chem., 62*(11), A633–642.

Huang, Z., Kawi, S., & Chiem, Y. C., (2004). Solubility of cholesterol and its esters in supercritical carbon dioxide with and without co-solvents. *Journal of Supercritical Fluids, 30,* 25–39.

Jafari, N. S., Abolghasemi, S. H., Moosavian, H., & Maragheh, M. A., (2010). Prediction of solute solubility in supercritical carbon dioxide: A novel semi-empirical model. *Chem. Eng. Res. Des., 8,* 893–898.

Johnston, K. P., Perk, D. G., & Kim, S., (1989). Modeling supercritical mixtures: How predictive is it? *Ind. Eng. Chem. Res., 26,* 1115–1125.

Jouyban, A., Chan, H. K., & Foster, N. R., (2002). Mathematical representation of solute solubility in supercritical carbon dioxide using empirical expressions. *J. Supercrit. Fluids, 24,* 19–35.

Khansary, M. A., Amiri, F., Hosseini, A., Sani, A. H., & Shahbeig, H., (2015). Representing solute solubility in supercritical carbon dioxide: A novel empirical model. *J. Chem. Eng. Research and Design, 93,* 355–365.

Kosal, E., Lee, C. H., & Holder, G. D., (1992). Solubility of progesterone, testosterone, and cholesterol in supercritical fluids. *Journal of Supercritical Fluids, 5*(3), 169–179.

Kumar, S. K., & Johnston, K. P., (1988). Modeling the solubility of solids in supercritical fluids with density as the independent variable. *J. Supercrit. Fluids, 1*(1), 15–22.

Kumovo, A. C., & Hassan, M., (2007). Supercritical carbon dioxide extraction of andrographolide from and *Rographis paniculata*: Effect of solvent flow rate, pressure and temperature. *Chin. J. Chem. Eng., 15*(6), 877–883.

McKinnon, I., & Parratt, J., (2002). *Organic Synthesis in Supercritical Fluids.* Press Release, Thomas Swan & Co. Ltd., UK.

Méndez-Santiago, J., & Teja, A. S., (2000). Solubility of solids in supercritical fluids: Consistency of data and a new model for co-solvent systems. *Ind. Eng. Chem. Res., 39,* 4767–4771.

Neves, G. B. M., (1996). Solubility of cholesterol and making butter oil in supercritical carbon dioxide. *M.Sc. Thesis,* State University of Campinas, Campinas-Brazil.

Norhuda, I., & Mohd, O. A. K., (2009). Mass transfer modeling in a packed bed of palm kernels under supercritical conditions. *International Journal of Chemical, Molecular, Nuclear, Materials and Metallurgical Engineering, 3*(1), 29–32.

Ouyang, L. B., (2011). New correlations for predicting the density and Viscosity of supercritical carbon dioxide under conditions expected in carbon capture and sequestration operations. *Journal of the Open Petroleum Engineering, 4*, 13–21.

Saldan, M. D., Homen, E. M., & Mohamed, R. S., (1997). Extraction of cholesterol with mixtures of carbon dioxide and supercritical ethane. *Sci. Technol. Aliment, 17*(4), 389–392.

Shishikura, A., Fujimoto, K., & Kaneda, T., (1996). Modification of butter oil by extraction with supercritical carbon dioxide. *Agric. Biol. Chem., 50*(5), 1209–1215.

Taylor, L. T., (1996). *Supercritical Fluid Extraction*. Wiley-Interscience Publication. John Wiley & Sons Inc., New York.

Vedaraman, N., Srinivasakannan, C., Brunner, G., Ramabrahma, B., & Rao, P. G., (2005). Experimental and modeling studies on extraction of cholesterol from cow brain using supercritical carbon dioxide. *Journal of Supercritical Fluids, 34*, 27–34.

Verma, D. K., Dhakane, J. P., Mahato, D. K., Billoria, S., Bhattacharjee, P., & Srivastav, P. P., (2018). Supercritical fluid extraction (SCFE) for rice aroma chemicals: Recent and advance extraction method. In: Verma, D. K., & Srivastav, P. P., (eds.), *Science and Technology of Aroma, Flavor and Fragrance in Rice.* Apple Academic Press, USA.

Wakao, N., & Kajuei, S., (1982). *Heat and Mass Transfer in Packed Beds* (p. 364). Gordon and Breach Science Publishers, USA.

Wang, X. R., Wu, Y., Wang, J. F., Dai, Y. P., & Xie, D. M., (2015). Thermo-economic analysis of a recompression supercritical $CO_2$ cycle combined with a transcritical $CO_2$ cycle. *Proceedings of ASME Turbo Expo 2015: Turbine Technical Conference and Exposition GT2015.* Montréal, Canad.

Yu, Z., Singh, B., Rizvi, S. S. H., & Zollewg, J. A., (1994). Solubilities of fatty acids, fatty acid esters, and fats and oils in supercritical carbon dioxide. *J. Supercrit. Fluids, 7*, 51–61.

Yun, S. L. J., Liong, K. K., Guardial, G. S., & Foster, N. R., (1991). Solubility of cholesterol in supercritical carbon dioxide. *Ind. Eng. Chem. Res., 30*, 2476–2482.

# CHAPTER 6

# Microwave-Convective Drying of Ultrasound Osmotically Dehydrated Tomatoes

JOÃO RENATO DE JESUS JUNQUEIRA,[1]
FRANCEMIR JOSÉ LOPES,[1] JEFFERSON LUIZ GOMES CORRÊA,[1]
KAMILLA SOARES DE MENDONÇA,[1] RANDAL COSTA RIBEIRO,[2] and
BRUNO ELYESER FONSECA[2]

[1]*Food Science Department, Federal University of Lavras, Lavras, Minas Gerais State, Brazil, E-mail: jrenatojesus@hotmail.com (J.R.D.J. Junqueira); E-mail: francemirlopes@yahoo.com.br (F.J. Lopes); E-mail: jefferson@dca. ufla.br (J.L.G. Corrêa); E-mail: keamendonca@msn.com (K.S.D. Mendonça)*

[2]*Engineering Department, Federal University of Lavras, Lavras, Minas Gerais State, Brazil, E-mail: randalribeiro8@gmail.com (R.C. Ribeiro); E-mail: brunoelyezerfonseca@yahoo.com.br (B.E. Fonseca)*

## 6.1 INTRODUCTION

Tomatoes are widely consumed all over the world, both fresh and in several processed products (Abbasi Souraki et al., 2014; Corrêa et al., 2016). Among the processed ones, dehydrated tomatoes have the advantages of convenience and high lycopene content (Purkayastha et al., 2011).

Drying is, perhaps, the oldest method for the conservation of food (Menezes et al., 2013; Verma et al., 2017). Although it carries out to stable products (Marques et al., 2014), it leads to physical, chemical, and nutritional changes (Corrêa et al., 2011; Mendonça et al., 2017; Verma et al., 2017). Due to this fact, the focus of most current drying studies is to limit the changes with respect to the fresh product (Purkayastha et al., 2011). The arrangement between microwave and convective drying (CD), consecutively (Pilli et al., 2008) or intermittently (Soysal, 2009; Junqueira et al.,

2017a), has improved sensory properties, appearance, color, and texture. Another benefit provided by the use of microwave in drying processes is the reduction of the drying time, thus decreasing power consumption (Balbay and Sahin, 2013). To ensure the best quality, osmotic dehydration (OD) has been used as a pretreatment in microwave drying (MWD) (Junqueira et al., 2016). OD can be conducted at atmospheric pressure (Silva et al., 2011), with vacuum pulse (Fante et al., 2011; Junqueira et al., 2017b; Viana et al., 2014) or by using ultrasound (Al-Hilphy et al., 2016) during the OD processes (Cárcel et al., 2012; Corrêa et al., 2015; Mendonça et al., 2015).

The aim of the chapter entitled "Microwave-convective drying of ultrasound osmotically dehydrated tomatoes" was to evaluate different methods of drying (convective and microwave-convective) and the effect of ultrasound-assisted osmotic dehydration (UAOD) on the drying kinetics and quality of semi-dried tomatoes. Diffusive and empirical models were tested for fitting the experimental kinetic drying.

## 6.2   MATERIAL AND METHODS

### 6.2.1   MATERIAL

The tomatoes (*Lycopersicon esculentum* Mill) CV. Carmen was purchased in the local market (Lavras, MG state, Brazil). The fruits were chosen visually by size, weight, color intensity, and uniform firmness, to be considered homogeneous.

The raw material was characterized according to its centesimal composition in relation to moisture content (*MC*), ash, lipids, proteins, fiber (AOAC, 2007) and carbohydrates, as follows: *MC* $19.0 \pm 0.2$ kg/100 kg dry basis (d.b.), lipids $0.13 \pm 0.01$ kg/100 kg; protein $0.03 \pm 0.00$ kg/100 kg; fibers $1.25 \pm 0.13$ kg/100 kg; ash $0.45 \pm 0.10$ kg/100 kg and carbohydrates $2.25 \pm 0.01$ kg/100 kg. The water activity was $0.996 \pm 0.001$ and the lycopene content $36.80 \pm 5.13$ mg/100 g.

#### 6.2.1.1   SAMPLE PREPARATION

The selected tomatoes were washed in tap water, cut into halves and had the seeds removed. The samples used in the experiments were obtained with the aid of a stainless steel mold. The fruits were flat slices with dimensions $4.0 \pm 0.2$ cm length, $2.0 \pm 0.2$ cm width, and the thickness of the fruit pulp, $0.8 \pm 0.2$ cm.

## 6.2.2 METHODS

### 6.2.2.1 ULTRASOUND-ASSISTED OSMOTIC DEHYDRATION (UAOD)

The samples were divided into two sets. The first one was dried without pretreatment. The other set was previously treated by UAOD. The treatment was conducted in 250 mL Erlenmeyer flasks immersed in the ultrasound bath (Unique brand, model USC 2850 A, ultrasound frequency 25 kHz, and powder intensity 8 kWm$^{-3}$) at 30°C (Garcia-Nogueira et al., 2010) in six replicates. The solution concentration was 10 kg NaCl/100 kg solution. The fruit: solution ratio was 1:4 (w/w) to avoid dilution effects by the osmotic process (Junqueira et al., 2017b). The ultrasound intensity was determined by a calorimetric method (Loning et al., 2002). During the experiments, the temperature was controlled, and the maximum increase in temperature due to the ultrasound waves was 2°C.

### 6.2.2.2 CONVECTIVE AND MICROWAVE-CONVECTIVE DRYING

The drying experiments were performed in convective and intermittent microwave-convective methods. The final MC was set 4.0 kg water/kg d.b. This MC provides a semi-dehydrated product with the main characteristics preserved. When the drying involved microwave, the experiments were divided by a convective-microwave pulse rate (PR) given by Equation (1). The on-off timings corresponded to interspersed periods of time with microwave application (on) followed by a period without microwave intermittent drying (off).

$$PR = \frac{t_{on} + t_{off}}{t_{on}} \tag{1}$$

where, $t_{on}$ corresponds to the microwave on time [s] and $t_{off}$ to the microwave off time [s].

The on-off timing used was as follow: PR1: Continuous MWD; PR2: MWD 30 s, CD 30 s; PR2.5: MWD 30 s, CD 45 s; PR3: MWD 30 s, CD 60 s; and PR4: MWD 30 s, CD 90 s.

The temperature and air velocity of the heated air used in convective and in the convective part of the microwave-convective drying were 40 or 60°C and 2.22 m s$^{-1}$, respectively.

### 6.2.2.3 EXPERIMENTAL SETUP

The microwave convective system (Figure 6.1) was adapted from Soysal (2009). The microwave chamber is a domestic microwave oven with nominal power of 1500 W (Electrolux MEC41 31 L, Brazil). A thermally insulated rectangular duct (56.0 cm wide x 34.0 cm high x 24.0 cm long) connected the microwave chamber and the blower. It was responsible for addition of air inside the microwave chamber. Electrical resistances were placed inside the duct to heat the air. The heated air entered the microwave chamber in a perforated opening (10.0 cm x 8.0 cm) in stainless steel located in the back wall of the chamber. An additional opening (10.0 cm x 8.0 cm) was provided in the upper wall of the chamber for air removal. For continuous sample mass measurement during the drying process, a digital balance (3100 g ± 0.001 g) was coupled to the turntable inside the microwave chamber. A programmable logic controller (PLC) was used to control the drying type: convective, microwave or intermittent microwave-convective drying. The temperature and air velocity were controlled by a proportional integral derivative (PID) and a solid relay. Inside the air duct, the temperature was measured with a platinum resistance sensor PT100. For presenting a uniform distribution of microwave energy throughout the material and, consequently, homogeneous drying, the plate with the samples was set to rotate at a constant velocity.

### 6.2.2.4 MICROWAVE OVEN POWER MEASUREMENT

The microwave power used in the experiments was determined by the test method "IMPI 2 – Liter" (Buffer, 1993). The test was performed with five replications and the average initial power microwave density was 7.8 W/g.

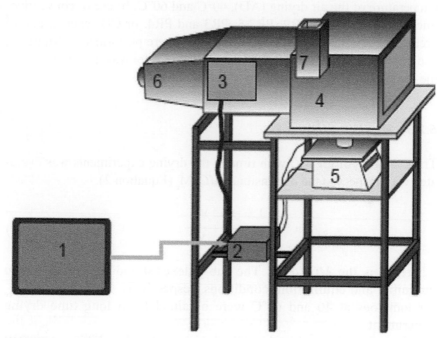

**FIGURE 6.1** Schematic view of the microwave convective dryer system. 1) Computer, 2) Magnetron, 3) Rectangular duct with internal resistances, 4) Microwave chamber, 5) Digital balance, 6) Blower, and 7) air exit.

### 6.2.2.5 QUALITY ANALYSES

Fresh, osmotically dehydrated and dried samples were analyzed with respect to *MC*, in a vacuum oven at 70°C according to the standard method 934.06 (AOAC, 2007), water activity at 25°C with a hygrometer (Aqualab, 3TE model, Decagon Devices Inc., Pullman, WA, USA), color parameters (Minolta CR 400 colorimeter, Minolta Camera Co. Ltd., Osaka, Japan) at 25°C according to the CIE *Lab* scale, and carotenoid content (Rodriguez-Amaya, 2001).

### 6.2.2.6 STATISTICAL ANALYSES

The drying experiments were performed in triplicate in a completely randomized factorial structure (2x6x2). The factors were as follows: a) temperature of the air drying (AD), 40°C and 60°C; b) use of convective-microwave PR: PR1, PR2, PR2.5, PR3 and PR4; or CD scheme; and c) pretreatment with UAOD. The experiments were performed in triplicate, and the Scott-Knott test at a 10% significance level was used to compare the means.

### 6.2.3   DRYING KINETICS

The evolution of the *MC* with time in the drying experiments was evaluated with respect to the dimensional *MC*, $M_r$ (Equation 2).

$$M_r = \frac{X - X_{eq}}{X_0 - X_{eq}}$$

(2)

where, X is the *MC* (d.b.). The sub-indexes 0 and eq correspond to the initial and equilibrium conditions, respectively. The values of $X_{eq}$ for tomatoes at 40 and 60°C were obtained for a long time drying experiment.

Most of the drying kinetics of biological materials exhibit exponential behavior (Corrêa et al., 2012). Therefore, two exponential models were tested: the diffusional Fick's model by the Crank's solution of the second Fick's Law (Crank, 1975) and the empirical Page's model (Isquierdo et al., 2013).

The Cranks solution of the second Fick's Law for an infinite plate is given by Equation (3) (Corrêa et al., 2010).

$$M_r = \left( \frac{8}{\pi^2} \sum_{i=1}^{\infty} \frac{1}{(2i+1)^2} \exp\left( -(2i+1)^2 \pi^2 D_{eff} \frac{t}{4L^2} \right) \right)$$

(3)

where, $D_{eff}$ is the effective diffusivity of the water [m²/s], i is the number of series terms and L is the characteristic length (sample half-thickness) [m] and t is the time [s].

The Page equation is presented by Equation (4).

$$M_r = \exp(-kt^n)$$

(4)

where, t is time [s], k is the drying rate [s$^{-1}$] and n is a parameter of adjustment.

## 6.3   RESULTS AND DISCUSSION

### 6.3.1   ULTRASOUND-ASSISTED OSMOTIC DEHYDRATION (UAOD)

The UAOD pretreatment reduced the *MC* from 19 to 9 kg/100 kg (d.b.). This reduction was due to the difference of chemical potential between the solution and the food in an osmotic process (Yadav and Singh, 2014). Moreover, the cavitation and the sponge effect promoted by the ultrasound waves could improve the mass transfer (Cárcel et al., 2012).

### 6.3.2   EXPERIMENTAL DRYING KINETICS

The drying time and the drying kinetics for the several tested conditions were presented in Table 6.1 and Figure 6.2, respectively.

**TABLE 6.1**   Drying Time (min) for Microwave-Convective Drying of Sliced Tomatoes

| Sliced Tomatoes | Temperature 40°C | | Temperature 60°C | |
|---|---|---|---|---|
| | Without UAOD | With UAOD | Without UAOD | With UAOD |
| PR1 | 6.84 | 21.2 | 6.96 | 21.0 |
| PR2 | 13.08 | 41.4 | 13.68 | 36.0 |
| PR2.5 | 16.08 | 51.0 | 16.44 | 42.6 |
| PR3 | 18.72 | 60.6 | 19.20 | 47.4 |
| PR4 | 24.12 | 70.2 | 26.88 | 78.0 |
| Convective | 654 | 1471 | 300 | 900 |

UAOD is ultrasound-assisted osmotic dehydration.

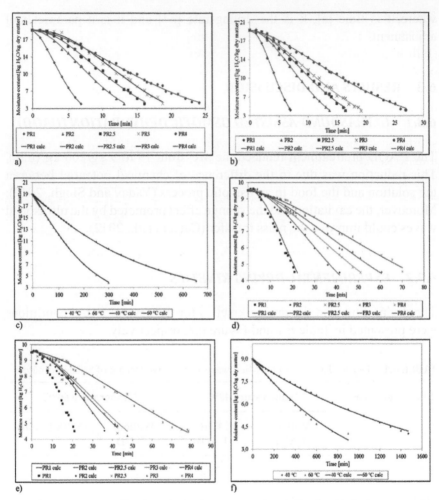

**FIGURE 6.2** Drying kinetics of sliced tomatoes. (a) 40°C, (b) 60°C, (c) convective drying, (d) with osmotic pretreatment, 40°C, (e) with osmotic pretreatment, 60°C, and (f) with osmotic pretreatment, convective drying. Calc is relative to the fit of the Page's model.

The results showed that the higher the microwave use in the microwave convective drying (lower PR), the shorter the drying time (Table 6.1, Figure 6.2), for both temperatures, with or without UAOD. The drying time for constant MWD (PR1) was on the order of 1% and 2% of the drying time of CD at 40°C and at 60°C, respectively. Microwave energy is converted to heat inside food due to the dipolar characteristic of the moisture of the food. The excitation of the water molecules with the microwave application reduces the internal resistance to drying (Chandrasekaran et al., 2013) with consequently higher drying rates and shorter drying time (Pilli et al., 2008). Moreover, the microwave application was more influent on drying time than the temperature. Consequently, the temperature, itself, does not presented significant influence on the microwave-convective drying time (Figure 6.2a, 6.2b and Figure 6.2d, 6.2e). On the other hand, for CD, it was observed that the increase from 40 to 60°C resulted in 50% shorter drying time (Figures 6.2c and 6.2f, Table 6.1). In a CD, the heat transfer occurs from the heated air to the material surface by convection and from the material surface to the inner part of the solid in a mainly diffusive way, by conduction.

The use of OD is suggested for improving quality characteristics (Corrêa et al., 2011) and reducing drying time (Askari et al., 2009; Junqueira et al., 2016). However, the drying time of the UAOD treated samples was higher than the ones of untreated samples (Table 6.1). Although the osmotic process provides a product with reduced $MC$, the incorporation of sodium chloride causes a reduction on the dielectric properties of the food (Al-Harahsheh et al., 2009). The dielectric properties reduction is directly proportional to the concentration of the salt (Al-Harahsheh et al., 2009; Nortemann et al., 1997). Moreover, the incorporation of salt reduces the free $MC$, difficulting the removal of the remaining water.

### 6.3.3  MATHEMATICAL MODELING

Among the several mathematical models for drying kinetics, Fick's diffusional equation and the semi-empirical Page model are commonly used for fitting MWD kinetics (Santos et al., 2010; Yu et al., 2015).

The Fick's model is based on the diffusional mechanism of a component with time in a preset dimension. The different drying methods developed in the present work are not complete diffusive processes, the diffusional mechanism in relevant in the transport of the moisture from the inner of the product to its surface (Yu et al., 2015). The Page's model is an empirical one with exponential characteristics. Due to the exponential behavior of the drying kinetics in microwave, microwave-convective, and CD, it presented adequacy for fitting in such cases (Balbay and Sahin, 2013), as presented in Figure 6.2.

The effective diffusivity ($D_{eff}$) from Fick's model and the k parameter from the Page's model could represent well the evolution of *MC* in the several tested modes (Table 6.2 and Table 6.3) showing coherent variations with microwave or convective aspect of the drying, temperature, and UAOD pre-treatment.

**TABLE 6.2**   Fit Parameters of the Fick Equation to the Drying Kinetics of Sliced Tomatoes

| Sliced Tomatoes | Temperature 40°C with UAOD | | | Temperature 40°C Without UAOD | | |
|---|---|---|---|---|---|---|
| | $D_{eff}$ x10$^8$ [m$^2$/s] | r$^2$ | %Var | $D_{eff}$ x10$^8$ [m$^2$/s$^1$] | r$^2$ | %Var |
| PR1 | 8.68 | 0.8929 | 81.68 | 0.896 | 0.8468 | 67.59 |
| PR2 | 3.76 | 0.8811 | 75.44 | 0.330 | 0.8154 | 60.83 |
| PR2.5 | 2.85 | 0.8743 | 72.70 | 0.201 | 0.8241 | 62.36 |
| PR3 | 2.28 | 0.9339 | 65.40 | 0.283 | 0.8957 | 64.85 |
| PR4 | 2.05 | 0.9114 | 77.31 | 0.201 | 0.8621 | 66.89 |
| Convective | 0.0878 | 0.9817 | 89.48 | 0.248 | 0.6188 | 41.43 |
| **Sliced Tomatoes** | **Temperature 60°C with UAOD** | | | **Temperature 60°C Without UAOD** | | |
| PR1 | 7.85 | 0.8488 | 74.07 | 0.943 | 0.8541 | 68.57 |
| PR2 | 3.56 | 0.8722 | 73.67 | 0.328 | 0.8122 | 61.64 |
| PR2.5 | 2.93 | 0.8869 | 74.68 | 0.355 | 0.8264 | 65.31 |
| PR3 | 2.66 | 0.9678 | 74.34 | 0.350 | 0.8264 | 63.77 |
| PR4 | 1.64 | 0.9184 | 78.43 | 0.257 | 0.8954 | 71.15 |
| Convective | 0.152 | 0.9643 | 81.93 | 0.023 | 0.9475 | 83.50 |

UAOD is ultrasound-assisted osmotic dehydration; $D_{eff}$ is the water diffusivity, r$^2$ is the correlation coefficient, %Var is the Explained variance.

**TABLE 6.3**   Fit Parameters of the Page Equation to the Drying Kinetics of Sliced Tomatoes

| Sliced Tomatoes | Temperature 40°C without UAOD | | | | Temperature 40°C with UAOD | | | |
|---|---|---|---|---|---|---|---|---|
| | k x 10$^3$ | n | r$^2$ | SE | k x 10$^3$ | n | r$^2$ | SE |
| PR1 | 7.69 | 1.57 | 0.9992 | 0.0031 | 1.98 | 1.95 | 0.9954 | 0.0088 |
| PR2 | 17.53 | 1.74 | 0.9977 | 0.0068 | 1.25 | 1.76 | 0.9894 | 0.0115 |
| PR2.5 | 8.98 | 1.84 | 0.9999 | 0.0095 | 1.50 | 1.60 | 0.9951 | 0.0088 |

**TABLE 6.3** *(Continued)*

| Sliced Tomatoes | Temperature 40°C without UAOD | | | | Temperature 40°C with UAOD | | | |
|---|---|---|---|---|---|---|---|---|
| | k x 10³ | n | r² | SE | k x 10³ | n | r² | SE |
| PR3 | 3.40 | 2.07 | 0.9999 | 0.0129 | 1.16 | 1.60 | 0.9927 | 0.0117 |
| PR4 | 8.30 | 1.63 | 0.9968 | 0.0107 | 1.05 | 1.54 | 0.9943 | 0.0093 |
| Convective | 2.02 | 1.02 | 0.9999 | 0.0018 | 0.74 | 0.95 | 0.9988 | 0.0174 |
| **Sliced Tomatoes** | **Temperature 60°C without UAOD** | | | | **Temperature 60°C with UAOD** | | | |
| PR1 | 36.22 | 1.94 | 0.9995 | 0.0029 | 1.27 | 1.79 | 0.9960 | 0.0104 |
| PR2 | 12.68 | 1.84 | 0.9983 | 0.0020 | 1.13 | 1.79 | 0.9893 | 0.0115 |
| PR2.5 | 10.25 | 1.77 | 0.9961 | 0.0103 | 2.14 | 1.57 | 0.9615 | 0.0226 |
| PR3 | 3.40 | 2.08 | 0.9999 | 0.0020 | 1.87 | 1.58 | 0.9519 | 0.0261 |
| PR4 | 8.30 | 1.56 | 0.9978 | 0.0094 | 1.07 | 1.52 | 0.9949 | 0.0111 |
| Convective | 2.37 | 1.14 | 0.9999 | 0.0019 | 0.92 | 1.02 | 0.996 | 0.0099 |

UAOD is ultrasound-assisted osmotic dehydration; $D_{eff}$ is the water diffusivity, n is the parameter of adjustment, r² is the correlation coefficient, and SE is the standard error.

## 6.3.4   QUALITY ASPECTS

### 6.3.4.1   COLOR

The color is changed with drying. The color parameters $a^*$ and $L^*$ of the dried product were related to the ones of the fresh product by the ratios $a^*/a^*_0$ and $L^*/L^*_0$ and the data are presented in Tables 6.4 and 6.5, respectively.

**TABLE 6.4**   $a^*/a^*_0$ of Sliced Tomatoes for Various Drying Conditions

| Sliced Tomatoes | Without UAOD | | With UAOD | |
|---|---|---|---|---|
| | 40°C | 60°C | 40°C | 60°C |
| Convective | 2.56 [a A] | 2.47 [b A] | 1.31 [b B] | 1.09 [c B] |
| PR4 | 1.43 [b C] | 2.99 [a A] | 1.22 [b C] | 1.86 [b B] |
| PR3 | 1.13 [b C] | 2.56 [b A] | 1.51 [b B] | 2.38 [a A] |
| PR2.5 | 1.42 [b C] | 2.47 [b A] | 1.89 [a B] | 1.59 [b C] |
| PR2 | 1.13 [b C] | 2.26 [b A] | 1.88 [a B] | 1.76 [b B] |
| PR1 | 1.35 [b B] | 1.58 [c B] | 1.95 [a A] | 1.48 [b B] |

UAOD means followed by the same letter do not differ significantly by Scott-Knott test at 10% probability, Lowercase in lines and uppercase in columns.

**TABLE 6.5** $L^*/L^*_0$ of Sliced Tomatoes for Various Drying Conditions

| Sliced Tomatoes | Without UAOD | | With UAOD | |
|---|---|---|---|---|
| | 40°C | 60°C | 40°C | 60°C |
| Convective | 0.84 [a C] | 1.21 [a A] | 0.75 [b D] | 0.92 [b B] |
| PR4 | 0.85 [a C] | 1.00 [b A] | 0.92 [a B] | 0.78 [c D] |
| PR3 | 0.74 [b C] | 0.88 [c A] | 0.91 [a A] | 0.82 [c B] |
| PR2.5 | 0.67 [c C] | 0.90 [c A] | 0.92 [a A] | 0.82 [c B] |
| PR2 | 0.62 [d B] | 0.88 [c A] | 0.93 [a A] | 0.90 [b A] |
| PR1 | 0.60 [d D] | 0.74 [d C] | 0.90 [a A] | 0.97 [a A] |

UAOD means followed by the same letter do not differ significantly by Scott-Knott test at 10% probability, lowercase in lines, and uppercase in columns.

According to the literature, the parameter $a^*$ could present value close to the fresh one when the use of osmotic treatment is employed (Zhao et al., 2013). In the present work, it was true for all drying processes at 60°C, for CD at 40°C and for the condition with the higher convective-microwave PR, PR4. For the other conditions at 40°C, the opposite was observed. It can be inferred that the osmotic treatment could diminish the influence of the temperature and the long-time exposition to the heated air. Within the same condition with respect to the use or not of osmotic pretreatment, the temperature did not present a clear trend. In general, within the same condition of drying actuation (convective or PR), the drying experiments at 40°C without UAOD presented the lowest values of $a^*/a^*_0$, with $a^*$ nearest to the fresh value, $a^*_0$. The ratio $L^*/L^*_0$ was close to one for treated samples dried at 40°C, and for untreated samples dried at 60°C, what means that large time process cause browning that could be reduced with the use of UAOD. With respect to the microwave use, it carried out to the values of $L^*/L^*_0$ close to the unit for pretreated samples. The opposite was observed for the untreated samples. In sum, the association of osmotic pretreatment with microwaves could reduce browning.

### 6.3.4.2   WATER ACTIVITY ($A_W$)

The $a_w$ of the samples submitted to drying are presented in Table 6.6.

**TABLE 6.6**  Water Activity for Different Drying Conditions

| Sliced Tomatoes | Without UAOD | | With UAOD | |
|---|---|---|---|---|
| | **40°C** | **60°C** | **40°C** | **60°C** |
| Convective | 0.966 [b B] | 0.988 [a A] | 0.942 [b C] | 0.964 [a B] |
| PR4 | 0.989 [a A] | 0.960 [c B] | 0.956 [a B] | 0.953 [b B] |
| PR3 | 0.960 [b A] | 0.958 [c A] | 0.959 [a A] | 0.960 [a A] |
| PR2.5 | 0.960 [b A] | 0.983 [a A] | 0.940 [b C] | 0.952 [b B] |
| PR2 | 0.962 [b B] | 0.982 [a A] | 0.960 [a B] | 0.942 [b C] |
| PR1 | 0.967 [b A] | 0.972 [b A] | 0.944 [b B] | 0.950 [b B] |

UAOD means followed by the same letter do not differ significantly by Scott-Knott test at 10% probability, lowercase in lines, and uppercase in columns.

The osmotic treatment was statistically relevant on $a_w$. The differences in drying methods and the temperature were not significative. The samples osmotically pretreated presented smaller $a_w$. Besides the water removal, the osmotic process involves solid gain (SG), represented in the present work by the incorporation of NaCl in the food structure (Viana et al., 2014). Both water removal and solid incorporation reduce the water availability, or the $a_w$.

### 6.3.4.3  LYCOPENE CONTENT

The results of the lycopene content of semi-dried tomato samples are shown in Table 6.7. The osmotic pretreatment with NaCl solution led to a reduced lycopene content. Temperature, in general, was not relevant. For untreated samples, the higher the PR, the higher the lycopene content. However, the highest values were observed for CD.

**TABLE 6.7**  Lycopene Content for Various Drying Conditions [mg/100 g]

| Sliced Tomatoes | Without UAOD | | With UAOD | |
|---|---|---|---|---|
| | **40°C** | **60°C** | **40°C** | **60°C** |
| Convective | 567.71 [a A] | 629.22 [a A] | 47.24 [b C] | 298.05 [a B] |
| PR4 | 118.92 [e B] | 225.54 [b A] | 91.38 [a B] | 120.20 [c B] |
| PR3 | 257.44 [d A] | 274.28 [b A] | 160.48 [a B] | 153.67 [b B] |
| PR2.5 | 214.19 [d A] | 143.47 [c B] | 123.39 [a B] | 28.98 [d C] |
| PR2 | 313.99 [c A] | 152.99 [c B] | 27.88 [b C] | 44.40 [d C] |
| PR1 | 390.97 [b A] | 173.62 [c B] | 24.90 [b C] | 179.46 [b B] |

UAOD means followed by the same letter do not differ significantly by Scott-Knott test at 10% probability, lowercase in lines, and uppercase in columns.

### 6.3.4.4  OPTIMAL DRYING CONDITION

Considering the drying time, color variation, final water activity, and lyco-pene content, the optimal drying condition was microwave-convective drying PR2 at 40°C without UAOD. This condition presented a drying time of 13 minutes, $a^*/a^*_0$ close to one, and $a_w$ statistically similar to the osmotic treated samples. Furthermore, this condition resulted in high lycopene content, statistically equivalent to that observed for CD.

## 6.4  SUMMARY AND CONCLUSION

The combined microwave-convective drying could carry out to a reduced time process with respect to the CD with the advantage of a relevant quality product. The drying of tomato slices by convective and intermit-tent microwave convective methods, pretreated or not by UAOD (25 kHz, 8 kWm$^{-3}$, 30°C, 10 kg NaCl/100 kg solution) was studied. The convective aspects were: air temperature 40°C and 60°C, and 2.22 m s$^{-1}$. Although the UAOD pretreatment reduced water activity, due to the salt incorpora-tion, samples without pretreatment presented higher lycopene content and water diffusivity during drying. Shorter drying time and higher diffusivity were obtained by increasing the microwave use and the temperature. The best fit of drying kinetics was verified for the Page's model in compared to Fick's model. Taking into account the quality and the drying aspects of semi-dried tomatoes, the best condition to process it is at convective-microwave pulse ratio of PR2, at 40°C, without UAOD pretreatment.

## KEYWORDS

- convective drying
- dielectric properties
- Fick's model
- food engineering
- heat transfer
- mass transfer

- **mathematical modeling**
- **microwave drying**
- **osmotic dehydration**
- **semi-dried fruits**
- **ultrasound waves**
- **water activity**

## REFERENCES

Abbasi, S. B., Ghavami, M., & Tondro, H., (2014). Comparison between continuous and discontinuous method of kinetic evaluation for osmotic dehydration of cherry tomato. *Journal of Food Processing and Preservation, 38*(6), 2167–2175.

Al-Harahsheh, M., Al-Muhtaseb, A. H., & Magee, T. R. A., (2009). Microwave drying kinetics of tomato pomace: Effect of osmotic dehydration. *Chemical Engineering and Processing: Process Intensification, 48*, 524–531.

Al-Hilphy, A. R. S., Verma, D. K., Niamah, A. K., Billoria, S., & Srivastav, P. P., (2016). Principles of ultrasonic technology for treatment of milk and milk products. In: Meghwal, M., & Goyal, M. R., (eds.), *Food Process Engineering: Emerging Trends in Research and Their Applications* (Vol. 5). As part of book series on "Innovations in Agricultural and Biological Engineering," Apple Academic Press, USA.

AOAC., (2007). *Association of Official Analytical Chemists: Official Methods of Analysis.* (18th edn.) Washington: DC.

Askari, G. R., Emam-Djomeh, Z., & Tahmasbi, M., (2009). Effect of various drying methods on texture and color of tomato halves. *Journal of Texture Studies, 40*, 371–389.

Balbay, A., & Şahin, Ö., (2013). Drying of pistachios by using a microwave assisted dryer. *Acta Scientiarum. Technology, 35*, 263–269.

Buffler, C. R., (1993). *Microwave Cooking and Processing: Engineering Fundamentals for the Food Scientist* (1st edn.) Van Nostrand Reinhold, New York, USA.

Cárcel, J. A., Benedito, J., & Mulet, A., (2012). Food process innovation through new technologies : Use of ultrasound. *Journal of Food Engineering, 110*, 200–207.

Chandrasekaran, S., Ramanathan, S., & Basak, T., (2013). Microwave food processing-A review. *Food Research International, 52*, 243–261.

Corrêa, J. L. G., Braga, A. M. P., Hochheim, M., & Silva, M. A., (2012). The influence of ethanol on the convective drying of unripe, ripe, and overripe bananas. *Drying Technology, 30*, 817–826.

Corrêa, J. L. G., Branquinho, D. E., & Mendonça, K. S., (2016). Pulsed vacuum osmotic dehydration of tomatoes: Sodium incorporation reduction and kinetics modeling. *Lebensmittel-Wissenschaft and Technologie / Food Science and Technology, 71*, 17–24.

Corrêa, J. L. G., Dev, S. R. S., Gariepy, Y., & Raghavan, G. S. V., (2011). Drying of pineapple by microwave-vacuum with osmotic pretreatment. *Drying Technology, 29*, 1556–1561.

Corrêa, J. L. G., Justus, A., Oliveira, L. F., & Alves, G. E., (2015). Osmotic dehydration of tomato assisted by ultrasound: Evaluation of the liquid media on mass transfer and product quality. *International Journal of Food Engineering, 11*, 505–516.

Corrêa, J. L. G., Pereira, L. M., Vieira, G. S., & Hubinger, M. D., (2010). Mass transfer kinetics of pulsed vacuum osmotic dehydration of guavas. *Journal of Food Engineering, 96*, 498–504.

Crank, J., (1975). *The Mathematics of Diffusion* (1st edn.) Clarendon Press: Oxford.

Fante, C., Corrêa, J., Natividade, M., Lima, J., & Lima, L., (2011). Drying of plums (*Prunus sp*, c.v Gulfblaze) treated with KCl in the field and subjected to pulsed vacuum osmotic dehydration. *International Journal of Food Science and Technology, 46*, 1080–1085.

Garcia-Noguera, J., Oliveira, F. I., Gallão, M. I., Weller, C. L., Rodrigues, S., & Fernandes, F. A., (2010). Ultrasound-assisted osmotic dehydration of strawberries: Effect of pretreatment time and ultrasonic frequency. *Drying Technology, 28*, 294–303.

Isquierdo, E., Borém, F., De Andrade, E., Correa, J., De Oliveira, P., & Alves, G., (2013). Drying kinetics and quality of natural coffee. *Transactions of the ASABE, 56*, 1003–1010.

Junqueira, J. R. J., Corrêa, J. L. G., & Ernesto, D. B., (2017a). Microwave, convective and intermittent microwave-convective drying of pulsed vacuum osmodehydrated pumpkin slices. *Journal of Food Processing and Preservation, 41*(6), e13250.

Junqueira, J. R. J., Corrêa, J. L. G., Mendonça, K. S., Resende, N. S., Vilas, B. E. V., & De, B., (2017b). Influence of sodium replacement and vacuum pulse on the osmotic dehydration of eggplant slices. *Innovative Food Science and Emerging Technologies, 41*, 10–18.

Junqueira, J. R. J., Mendonça, K. S., & Corrêa, J. L. G., (2016). Microwave drying of sweet potato (*Ipomoea batatas* (L.). Slices: Influence of the osmotic pretreatment. *Defect and Diffusion Forum, 367*, 167–174.

Löning, J. M., Horst, C., & Hoffmann, U., (2002). Investigations on the energy conversion in sonochemical processes. *Ultrasonics Sonochemistry, 9*, 169–179.

Marques, G. R., Borges, S. V., Botrel, D. A., Costa, J. M. G. D., Silva, E. K., & Corrêa, J. L. G., (2014). Spray drying of green corn pulp. *Drying Technology, 32*, 861–868.

Mendonça, K. S., Corrêa, J. L. G., Jesus, J. R., Pereira, M. C. A., & Vilela, B. M., (2015). Optimization of osmotic dehydration of yacon slices. *Drying Technology, 34*(4), 386–394.

Mendonça, K. S., Corrêa, J. L. G., Junqueira, J. R. J., Cirillo, M. A., Figueira, F. V., & Carvalho, E. E. N., (2017). Influences of convective and vacuum drying on the quality attributes of osmo-dried pequi (*Caryocar brasiliense* Camb.) slices. *Food Chemistry, 224*, 212–218.

Menezes, M. L., Kunz, C. C., Perine, P., Pereira, N. C., Dos, S. O. A. A., De Barros, S. T. D., (2013). Analysis of convective drying kinetics of yellow passion fruit bagasse. *Acta Scientiarum. Technology, 35*, 291–298.

Nortemann, K., Hilland, J., & Kaatze, U., (1997). Dielectric properties of aqueous NaCl solutions. *Journal of Physical Chemistry A, 101*, 6864–6869.

Pilli, T., Lovino, R., Maenza, S., Derossi, A., & Severini, C., (2008). Study on operating conditions of orange drying processing: Comparison between conventional and combined treatment. *Journal of Food Processing and Preservation, 32*, 751–769.

Purkayastha, M. D., Nath, A., Deka, B. C., & Mahanta, C. L., (2011). Thin layer drying of tomato slices. *Journal of Food Science and Technology, 50*, 642–653.

Rodriguez-Amaya, D. B., (2001). *A Guide to Carotenoid Analysis in Foods* (1ˢᵗ edn.). Washington DC: ILSI Press.

Santos, C. T., Bonomo, R. F., Chaves, M. A., Fontan, R. D. C. I., & Bonomo, P., (2010). Cinética e modelagem da secagem de carambola (*Averrhoa carambola* L.) em secador de bandeja. *Acta Scientiarum. Technology, 32*, 309–313.

Silva, M. A. C., Corrêa, J. L. G., & Silva, Z. E. D. A., (2011). Drying kinetics of West Indian cherry : Influence of osmotic pretreatment. *Boletim do CEPPA, 29*, 193–202.

Soysal, Y., (2009). Intermittent microwave-convective drying of red pepper: Drying kinetics, physical (color and texture) and sensory quality. *Biosystems Engineering, 103*, 455–463.

Verma, D. K., Kapri, M., Billoria, S., Mahato, D. K., & Prem, P. S., (2017). Effects of thermal processing on nutritional composition of green leafy vegetables: A review. In: Verma, D. K., & Goyal, M. R., (eds.), *Engineering Interventions in Foods and Plants*. As part of book series on "Innovations in Agricultural and Biological Engineering." Apple Academic Press, USA.

Viana, A. D., Corrêa, J. L. G., & Justus, A., (2014). Optimization of the pulsed vacuum osmotic dehydration of cladodes of fodder palm. *International Journal of Food Science and Technology, 49*, 726–732.

Yadav, A. K., & Singh, S. V., (2014). Osmotic dehydration of fruits and vegetables: A review. *Journal of Food Science and Technology, 51*, 1654–1673.

Yu, H. M., Zuo, C. C., & Xie, Q. J., (2015). Drying characteristics and model of Chinese hawthorn using microwave coupled with hot air. *Mathematical Problems in Engineering*, 1–15.

Yu, X., Schmidt, A. R., & Schmidt, S. J., (2009). Uncertainty analysis of hygrometer-obtained water activity measurements of saturated salt slurries and food materials. *Food Chemistry, 115*, 214–226.

Zhao, D., Zhao, C., Tao, H., An, K., Ding, S., & Wang, Z., (2013). The effect of osmosis pretreatment on hot-air drying and microwave drying characteristics of chili (*Capsicum annuum* L.) flesh. *International Journal of Food Science and Technology, 48*, 1589–1595.

# CHAPTER 7

# Ultrasound-Assisted Osmotic Dehydration in Food Processing: A Review

VAHID MOHAMMADPOUR KARIZAKI

*Chemical Engineering Department, Quchan University of Advanced Technology, Quchan, Iran, E-mail: mohammadpour_vahid@yahoo.com*

## 7.1 INTRODUCTION

Almost all types of foodstuffs in our daily diet require some kind of preservation to minimize or stop chemical damage and microbial spoilage (Jangam, 2011; Verma and Srivastav, 2017). The various studies have addressed the different methods of food preservation such as canning, salting, vacuum packaging, freezing, irradiation, and drying (Chouliara et al., 2004; Jiang et al., 2017; Kung et al., 2017; Kachele et al., 2017; Mahato et al., 2017; Rajkumar et al., 2017). Among these, drying is one of the oldest and the most commonly used techniques of food preservation (Jangam, 2011). Many diverse varieties of food materials such as fruits, vegetables, herbs, spices, meat, and marine products, grains, and dairy products can be routinely preserved using drying (Defraeye, 2017; Defraeye and Verboven, 2017; Tontul and Topuz, 2017).

Drying mainly consists of moisture removal from raw material due to the simultaneous heat and mass transfer operation (Onwude et al., 2016; Verma et al., 2017). The MC (MC) of food materials has a wide range from 25–30% in grains to 90% or even more in some vegetables and fruits (Jangam, 2011). The choice of suitable dryer type for each food product depends on several factors such as initial moisture content (MC) of raw material, size, shape, and form of the product (Mujumdar 2004). [A comprehensive investigation on the different types of drying methods and also various kinds of products that may be dried are found in Tadeusz and Mujumdar (2009) and Mujumdar (2014)].

Osmotic dehydration (OD) is a simple and inexpensive method for moisture removal that has been applied for numerous food products (Toğrul and İspir, 2008; García-Segovia et al., 2010; Seguí et al., 2012; Silva et al., 2012; Abraão et al., 2013; Porciuncula et al., 2013; Rastogi et al., 2014). It is also followed by other processes such as freeze drying (FD), air drying (AD), and deep fat frying to obtain a product with higher quality (Ahmed et al., 2016; Da Costa Ribeiro et al., 2016; Sette et al., 2016; Cano-Lamadrid et al., 2017; Prosapio and Norton, 2017). It is shown that the use of this dehydration method as a pretreatment before other unit operations can lead to an increase in the sensory, nutritional, and functional properties of food products (Ciurzyńska et al., 2016). However, OD is a time-consuming process, and therefore a new way to increase the efficiency of the operation is required (Kowalski et al., 2015). There are several pretreatments used to increase the drying rate, including thawing, freezing, extrusion, pinning, abrasion, or drilling holes on the product surface, hot water blanching (HWB), steam blanching (SB), $SO_2$ treatment. Also, to enhance mass transfer rate during the dehydration process, several techniques have been applied such as microwave, radiofrequency, ultraviolet (UV), infrared, ohmic, pulsed electric fields (PEF), pulsed-vacuum, supercritical, high pressure, rotation, and ultrasound (Xu et al., 2014; Onwude et al., 2016). Among these, ultrasound technology in food processing is relatively new (Duan et al., 2008; Al-Hilphy et al., 2016). However, combination of ultrasound technique with OD method has recently gained a considerable amount of interest with several researches on applications, drying kinetics, and products (Duan et al., 2008; Fernandes and Rodrigues, 2008; Onwude et al., 2016; Amami et al., 2017; Barman and Badwaik, 2017; Corrêa et al., 2017).

The aim of the chapter entitled "ultrasound-assisted osmotic dehydration (UAOD): a review" to focused on the application of UAOD in food industry. The recent advances and developments on this subject are reviewed.

## 7.2  ULTRASOUND-ASSISTED OSMOTIC DEHYDRATION (UAOD)

### 7.2.1  OSMOTIC DEHYDRATION (OD)

OD can preserve nutritional attributes and provide an extension of shelf-life (SL) of various fruits and vegetables (Amami et al., 2017).

Consumption of osmotically dehydrated products as a part of foodstuffs has gained a noticeable increase. OD is a counter-current mass transfer operation, in which the osmotic agents (e.g., glucose, sucrose, fructose, and sodium chloride that are used according to the food material) flow into the sample, while moisture is removed from the interior of the sample to the hypertonic solution (HS) (Ahmed et al., 2016).

The osmotic pressure difference between the HS and the food tissue provides the required driving force for removing moisture from food material to the osmotic solution. The mass transfer between the HS and the foodstuff may have an effect on the yield and the quality of the dehydrated fruits and vegetables (Ahmed et al., 2016). The osmotically dehydrated fruits and vegetables can be used for direct usage, as well as used in products such as desserts, yogurt, ice cream, and confectionery materials.

The physicochemical characteristic of food materials such as color, texture, rehydration capacity, and flavor can be affected during the dehydration process (Amami et al., 2017). The findings show that the OD is generally performed at mild temperatures (between 20 to 60°C). It takes from 30 minutes to 5 hours or even more depending on the food material that is being processed, its initial and final MC, the concentration and the type of the osmotic agents, and the temperature of the osmotic solution (Fernandes and Rodrigues 2008). Also, the rate of moisture removal depends on the other parameters, such as size and shape of the sample, agitation type and speed, the solution to food pieces ratio, and the vacuum pressure, if applied (Torreggiani and Bertolo, 2004).

Water loss (WL) and solids gain (SG) induced by OD also result in changing the microstructure of plant tissues, such as deformation of cell walls, cell collapse, and rupture of cellular bonds (Deng and Zhao, 2008). The effect of OD on the microstructure and texture of food materials have been presented in reported studies (Del Valle et al., 1998; Prothon et al., 2001; Castelló et al., 2009). OD can be used as an independent and sole dehydration way in a drying process, or can be applied as a pre-treatment before food processing. It is the most reported and popular pre-treatment used prior to conventional air-drying (Fernandes et al., 2008).

The complex cellular structure of food materials acts as a semi-permeable membrane in the OD process (Rodrigues et al., 2009). Since this pressure difference is a sole driving force for exchanging mass during the process, OD is a time-consuming and low efficient process. For increasing processing efficiency and enhancement of final product quality, several ways have been recommended by investigators (Corrêa et al., 2010; Wang et al., 2013; Patel and Sutar, 2016). In recent years, the combination of OD with other techniques has considerably been increased in order to improve food quality parameters or speed up the dehydration time (Jangam, 2011; Amami et al., 2017).

Summarizing the available scientific literature, the following techniques may be carried out for OD assistance: microwave, blanching, pulsed electric fields, high hydrostatic pressure, centrifugal force (CF), and ultrasonic waves (UW) (Moreno et al., 2000; Erle and Schubert, 2001; Ade-Omowaye et al., 2003; Rastogi et al., 2006; Botha et al., 2012; Goula and Lazarides, 2012; Parniakov et al., 2016). Using these techniques accompanying with osmotic pretreatment before other processes result in better sensory acceptability and nutritional quality.

### 7.2.2   USING OF ULTRASOUND DURING DRYING PROCESS

In recent years, the use of the ultrasonic techniques in the food industry has been grown, and many applications of this technology have been considered relating to the use of UW to directly affect the product or process (Fernandes and Rodrigues, 2008; Al-Hilphy et al., 2016). Many researchers have used UW for increasing the drying rate of food materials. The acoustically assisted conventional AD allows the using of lower temperatures and may be beneficial for the drying of heat-sensitive products (Gallego-Juarez et al., 1999). Thus, the application of ultrasonic energy may positively contribute during the drying process. However, the use of this technology has been limited, due to the insufficient developments and/or the practical difficulties related to the needs of large-scale production.

The idea of using ultrasonic waves for the drying process is not new. The first experiments on the use of this technique in the drying operation were conducted in 1959 by Boucher and 1963 by Greguss, both the sonic and ultrasonic period. In those times, the processes of sonic and ultrasonic drying were applied mainly in the Soviet Union. The obtained finding of the 1950s demonstrated that the use of ultrasound accelerated the drying process without leading to a drastic enhancement in the temperature. This is why ultrasound waves were especially considered to increase the drying of temperature-sensitive products, such as foodstuffs (Musielak et al., 2016).

The current applications of ultrasound techniques in the food industry include extraction, processing, and preservation. The preservation using UW has been the subject of several research and development projects around the world (Mason et al., 2003; Barbosa-Cánovas and Bermúdez-Aguirre, 2010; Chemat et al., 2011). The schematic diagram of Figure 7.1 shows how ultrasound technology is applied as assistance for the OD process. The main purpose of this combination is creation an effective way for the preservation of foodstuffs.

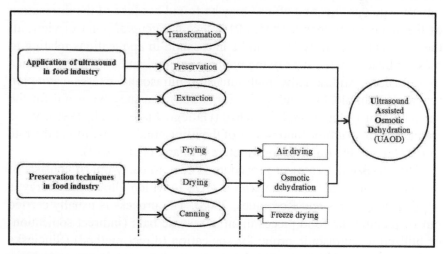

**FIGURE 7.1**   Combination of ultrasound technique with osmotic dehydration for creating an effective method of preservation.

Losses of some constituents, time-consuming procedures, low production yield, and high energy consumption may be occurred using the conventional OD. These shortcomings have resulted in the use of modern, sustainable, and green techniques in processing, which generally involve less energy and time, such as the ultrasound-assisted method (Chemat et al., 2011). The increasing number of relevant researches show that ultrasound is a very important and promising technology applied to help the OD (Yao, 2016). When UW (usually in a frequency range of 20–100 KHz) are applied to the food material to be dried, it diffuses into the solid tissue resulting in a rapid series of alternative compressions and expansions. The process is similar to a sponge which it is squeezed and released frequently (sponge effect) (Gallego-Juarez et al., 1999; Amami et al., 2017). The repeated stress makes dewatering simple and easy by developing microscopic channels for moisture removal. These channels may be utilized by water molecules as an appropriate and preferential pathway for diffusing toward the surface of the food material. As a result, the moisture effective diffusivity during the process is increased. The UW also decrease the diffusion boundary layer, as well as the convective mass transfer in the food piece is increased (Fernandes and Rodrigues, 2008). The increasing mass transfer is due to the oscillatory motion of an ultrasonic wave that causes acoustic streaming (Deng and Zhao, 2008). Also, the other subsequent impacts such as "heating effect" may be found because of the dissipation of the mechanical energy (Yao, 2016). From a general point of view, all these effects created by UW can be interesting in applications relating to mass or heat transfer, decreasing both the internal and external resistance to diffusion. Additionally, high-intensity ultrasound causes cavitation (expansion, implosion, and cavity formation) that may be useful for the separation of water strongly attached (Gallego-Juarez et al., 1999). Wave frequency, drying time, and power of the ultrasound greatly affect the rate of the OD and the final food's structure (Amami et al., 2017).

The ultrasonic treatment procedure is easy to carry out. It consists of the immersion of the food material in a HS to which ultrasound energy is applied (Fernandes and Rodrigues, 2008). The process is usually carried out by placing the food pieces in an ultrasonic bath (indirect sonication) or utilized an ultrasonic probe for sonication (direct method) (Siucińska and Konopacka, 2014). Figure 7.2 shows these two different techniques of sonication.

The operation should be taken for at least 10 min. Some investigators showed that the effect of the ultrasound at lower times is probably insignificant. The differences between indirect and direct sonication combined with OD of guava slices were considered by Kek et al. (2013). The authors showed that the direct method was more intensive in comparison with the indirect sonication. In addition, UW imposed indirectly on the samples led to high SG and WL with acceptable color change (Kek et al., 2013; Siucińska and Konopacka, 2014).

The use of UAOD is especially suggested for those fruits and vegetables that are sensitive to thermal treatment (e.g., strawberry and kiwifruit). The drying process utilizes ultrasonic energy, also known as acoustic drying that is a safe, beneficial, and environmentally friendly method (Siucińska and Konopacka, 2014).

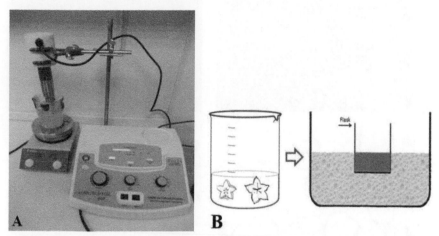

**FIGURE 7.2**   Schematic diagram of ultrasound-assisted osmotic dehydration. A) Instrument UAOD used by Karizaki and co-workers in 2013 for potato in which they applied direct sonication by probe for performing the process (Karizaki et al., 2013), and B) Used by Barman and Badwaik (2017) for carambola pieces presented in which they utilized indirect sonication in an ultrasonic bath for fruit processing.

However, this technique is not recommended when removing of a large amount of moisture from food material are required, because osmotic agent concentrating on the surface of the foodstuff complicates the water-elimination process (Siucińska and Konopacka, 2014). Color changes in a food material can impact on the overall acceptability of the food for consumers. Several investigators have reported the effect of UW on the color of fruits and vegetables during OD (Pingret et al., 2013). The UAOD process changes the food texture, especially because of breakdown of cells and pectin dissolution (Rodrigues et al., 2009).

As mentioned earlier, the sponge effect caused by sonication may be responsible for the enlargement of cell interspaces and the development of microscopic channels. Several micrographs of food tissues treated with UAOD are shown in Figure 7.3. These pictures show how a UAOD process results in the cell breakdown, and the enlargement of cell interspace.

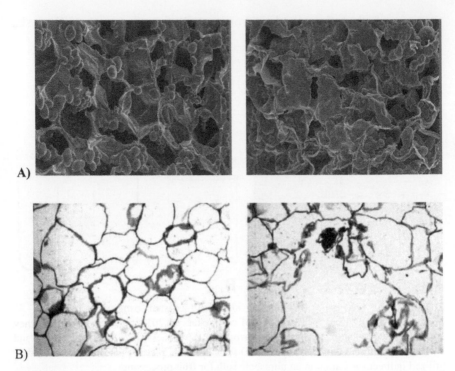

**FIGURE 7.3**    The microstructural changes in food tissues during the ultrasound-assisted osmotic dehydration process. (A) The left image shows the micrograph of fresh potato with normal morphology and uniform cell structure. The round-shaped and intact granules of starch are observed before UAOD treatment. The right image shows how UAOD treatment has ruptured and mechanically damaged the potato cells (Reprinted from Karizaki et al., 2013). (B) Micrograph of papaya piece after 30 min of UAOD (Reprinted from Rodrigues et al., 2009): The left image shows the medium size cell interspaces, and the right image shows papaya tissue with severe cell breakdown.

As a general result, based on the published reports, higher UAOD time of the food tissues, the higher degree of cell breakdown, and the larger microscopic channels (Rodrigues et al., 2009). The more disruption of the cell structure may cause to increase in the WL during UAOD in comparison with the conventional OD. Due to UAOD stress on the food structure, the state of the cell membrane may be changed from partial to total permeability, causing to important changes in tissue architecture (Deng and Zhao, 2008). The ability of ultrasound waves for interacting with the cellular structure of the food tissue creates new possibilities for pretreated product modifications, which impact the effectiveness of the diffusion mechanism (Siucińska and Konopacka, 2014).

Summarizing the findings of researchers in the literature, when applying UW for food material pretreatment prior drying the following modification may be included: higher effective moisture diffusivity, cellular structure changes (e.g., creation of microscopic channels), increase of WL, better color preservation, and higher drying rates (Siucińska and Konopacka, 2014).

Among the foodstuffs selected to UAOD or ultrasound pretreatment, special attention to fruits and vegetables has been paid by investigators. A summary of researches directly dealt with the UAOD is listed in Table 7.1.

Also, several researchers have considered the effect of UAOD prior to deep frying. Deep fat frying process is widely used for domestic consumption, as well as for industrial purposes. Due to the need for healthy foodstuffs of high quality, it is vital to find methods which can help decrease oil uptake (OU) in fried foods. High consumption of fried products especially in modern societies increase the risk of diseases such as hypertension, cardiovascular, cancer, obesity, and diabetes (Dehghannya et al., 2016).

The effect of UAOD prior to frying of potato is investigated by Karizaki et al. (2013). The authors studied the effects of the pretreatment on OU, MC, texture, color, and microstructure of the fried potatoes. They found that the UAOD process improve the quality of final product through reducing moisture and oil contents in comparison with the untreated samples (Karizaki et al., 2013; Dehghannya et al., 2016). Generally, it has been shown that reducing the initial MC of a food material decreases OU in the final fried product.

## 7.3   MODELING OF THE PROCESS

Drying process is a complex unit operation, which generally consists of simultaneous phenomena of mass and heat transfer (Amami et al., 2017). The complexity of the UAOD will be increased due to the effect of UW on the process, and the countercurrent diffusion of water/solute between osmotic solution and the sample. In the last few years, various models of drying operation have been presented through different phenomenological, fundamental, experimental, and kinematical ways. Some diffusion models recommend determination the moisture effective diffusivity based on the

**TABLE 7.1** A Summary of Reports on Osmotic Dehydration Combined with Ultrasound Waves

| Food Materials | Experimental Conditions | | | | | | | Results and Concluding Remarks | References |
|---|---|---|---|---|---|---|---|---|---|
| | Osmotic Dehydration | | | | Sonication | | | | |
| | Solution Temperature (°C) | Concentration | Osmotic Agent(s) | Ultrasound Time | Sonication Frequency | Intensity | | | |
| Apple | Room temperature | 60% (w/w) | Fructose | 180 | - | - | | The effects of osmotic dehydration combined with sonication on mechanical behavior, glass transition, and microstructure of freeze-dried and hot air-dried apples were evaluated. SEM pictures showed a larger amount of solid gain in the cells of ultrasound treated apples. | Deng and Zhao, 2008a, b |
| | 40–70 | 70°Brix | Sucrose | | 50 | - | | A significant increase in moisture and solute transport rates was observed when sonication was applied instead of agitation. | Simal et al., 1998 |

**TABLE 7.1** *(Continued)*

| | | | | | | | | |
|---|---|---|---|---|---|---|---|---|
| Banana, genipap, jambo, melon, papaya, pineapple, pinha, sapota | 50 | 25–70° | Food grade sucrose | 10–45 | 25 | 4000 | The moisture effective diffusivity of water in the samples is increased after the application of ultrasound waves; those results in reducing time of air-drying. | Fernandes and Rodrigues, 2008 |
| Blueberry | 21 | 55°Brix | Commercial sucrose | 180 | 850 | - | Ultrasound waves had a negative effect on anthocyanins and phenolics. | Stojanovic and Silva, 2007 |
| Broccoli | 35 | 40% (w/w) | Trehalose | 10–40 | 40 | 3300 | Compared with the conventional osmotic dehydration process with 2 h, UAOD with lower time (30 min) can get the higher value of glass transition temperature by increased the water loss and trehalose accumulation and reduction of the mobility of moisture in the broccoli cell tissue. | Xin et al., 2013 |

**TABLE 7.1** (Continued)

| | | | | | | | | |
|---|---|---|---|---|---|---|---|---|
| Carambola | 40–60 | 50–70° | Glucose, sucrose, fructose, glycerol | 10–30 | 25 | 4870 | With increasing in time of the UAOD process, the rehydration ratio and water loss were increased while texture and solid gain were decreased. | Barman and Badwaik, 2017 |
| Carrot | 30 | 40, 60% (w/w) | Fructose | 120 | 38 | - | It was observed that the combination of UAOD with intermittent-convective drying accelerates the rate of drying and improves the quality of the dried products. | Kowalski et al., 2015 |
| Cherry | 28 | 60% (w/w) | Glucose | 30 | 25 | - | It was found that intermittent drying of cherry by using the UAOD process contributes to lower drying time, smaller water activity, and better color preservation. | Kowalski and Szadzińska, 2014 |

**TABLE 7.1** (Continued)

| | | | | | | Remarks | Reference |
|---|---|---|---|---|---|---|---|
| Guava | 30 | 0–70° | Commercial sugar | Bath: 20–60 Probe: 6–20 | 25 | – | Using the UAOD process prior to air drying shortened the drying time by 33%, decreased the color change by 38%, and increased the moisture effective diffusivity by 35%. | Kek et al., 2013 |
| Kiwifruit | 25 | 61.5% (w/w) | Sucrose | 10–30 | 35 | – | Sonication alone had a negative effect on texture parameter. When combined with osmotic dehydration, ultrasound wave was able to maintain and improve the product characteristics.<br>**It was observed that ultrasound wave's treatment carried out for more than ten minutes had a positive impact on the mass exchange during osmotic dehydration. | Nowacka et al., 2017<br>Nowacka et al., 2014 |

**TABLE 7.1** (Continued)

| | | | | | | | | |
|---|---|---|---|---|---|---|---|---|
| Melon | 30 | 50% (w/w) | Sucrose | 10–30 | 25 | 4870 | The use of UAOD as a pretreatment could enhance the efficiency of air drying of melon fruit. | Dias da Silva et al., 2016 |
| Melon* | 42.5 | 70° | Food grade sucrose | 20, 30 | 25 | 4870 | Ultrasound waves increased moisture effective diffusivity due to the creation of microscopic channels. Osmotic dehydration, when conducted for less than 30 min reduces the moisture effective diffusivity due to the incorporation of an osmotic agent, but increases diffusivity when conducted for more than one hour due to the breakdown of sample cells decreasing the resistance to moisture diffusion. | Fernandes et al., 2008 |

**TABLE 7.1** (Continued)

| | | | | | | | | |
|---|---|---|---|---|---|---|---|---|
| Papaya | 30 | 25° | Food grade sucrose | 10–30 | 25 | 4870 | UAOD caused to a gradual distortion in the cells shape, loss of cellular adhesion, and the creation of large channels due to the rupture of the cell walls. | Rodrigues et al., 2009 |
| Potato | 25–65 | 15, 50% (w/w) | Sucrose, salt | 10–30 | 20 | - | UAOD was shown to have the advantage of the improvement the color of fried potatoes. | Karizaki et al., 2013 |
| Radish | 30 | 60% (w/w) | Food grade sucrose | 0–300 | 40 | 2890 | Sonication significantly decreased the dehydration time and subsequent freezing time. | Xu et al., 2014 |
| Sea cucumber | 24 | 20% (w/w) | Salt | 0–45 | 25 | - | It was found that UAOD can decrease by about 2 h the time required for microwave freeze drying of sea cucumber. | Duan et al., 2008 |
| Strawberry | 20–40 | 32.5, 65°Brix | Commercial sugar | 10–30 | 40 | 2000 | The effects of UAOD under the optimized operating conditions were considered on kinetics of air drying of strawberry. | Amami et al., 2017 |

**TABLE 7.1** *(Continued)*

| 30 | 25–50% (w/w) | Food grade sucrose | 10–45 | 0–40 | - | Creation of micro-channels through sonication and impact of osmotic pressure difference were found to be responsible for decreasing dehydration time of strawberry halves. | Garcia-Noguera et al., 2010 |

*The effects of ultrasound and osmotic dehydration on melon drying are investigated separately.

**The fruit pieces were treated with ultrasound waves before osmotic dehydration, not with a simultaneous process of sonication and osmotic dehydration.

second Fick's law (Seth and Sarkar, 2004; Ramos et al., 2010; Khir et al., 2011; Ruiz-Lópeza et al., 2012).

Response surface methodology (RSM) is an important statistical technique in the operation optimization (Amami et al., 2017). It is employed by researchers to study and optimize the induction of the main process parameters such as concentration of osmotic solution, ultrasound time, and the operation temperature (independent parameters) on the WL, and SG (dependent parameters).

Amami et al. (2017) applied a RSM model to the experimental data during the UAOD of strawberry. They investigated and optimized the effect of independent parameters on the weight reduction (WR), WL, and SG. Also, the authors calculated the activation energy and effective moisture diffusivity from analogous Arrhenius equation and similar-Fick's law, respectively (Amami et al., 2017). Finding and understanding the UAOD process models, such as mass transfer rates, mass transfer parameters (effective moisture diffusivity, mass Biot number, mass transfer coefficient), heat transfer parameters, textural, and structural relations associated with different conditions are very important for an appropriate design of a dehydration system (Deng and Zhao, 2008).

Also, mass and heat transfer parameters are beneficial and vital in understanding and description of the unit operations dynamic and predicting the rate of heat and mass transfer during food processes (Mosavian and Karizaki, 2012; Karizaki 2016). Further studies on modeling and determination of mass and heat transfer parameters during UAOD are required.

## 7.4   SUMMARY AND CONCLUSION

The usage of ultrasound is found to have an advantage of the enhancement of the efficiency and speed of OD. UAOD can be a good alternative to OD in food products as long as the decreasing in processing time, and color improvement are expected. The UAOD process was introduced, as well as a summary of reports on this technique for processing of foodstuffs was reviewed. From the investigation presented in the current chapter, this conclusion can be stated that the ultrasound treatment is beneficial for the removal of more moisture from fruit and vegetable tissue.

## KEYWORDS

- **dehydration**
- **food quality**
- **mass transfer**
- **moisture content**
- **osmosis**
- **osmotic dehydration**
- **preservation**
- **rehydration**
- **shelf-life**
- **sonication**
- **sponge effect**
- **ultrasonic waves**
- **ultrasound**
- **water loss**

## REFERENCES

Abraão, A. S., Lemos, A. M., Vilela, A., Sousa, J. M., & Nunes, F. M., (2013). Influence of osmotic dehydration process parameters on the quality of candied pumpkins. *Food and Bioproducts Processing, 91*(4), 481–494.

Ade-Omowaye, B. I. O., Rastogi, N. K., Angersbach, A., & Knorr, D., (2003). Combined effects of pulsed electric field pre-treatment and partial osmotic dehydration on air drying behavior of red bell pepper. *Journal of Food Engineering, 60*(1), 89–98.

Ahmed, I., Qazi, I. M., & Jamal, S., (2016). Developments in osmotic dehydration technique for the preservation of fruits and vegetables. *Innovative Food Science and Emerging Technologies, 34,* 29–43.

Al-Hilphy, A. R. S., Verma, D. K., Niamah, A. K., Billoria, S., & Srivastav, P. P., (2016). Principles of ultrasonic technology for treatment of milk and milk products. In: Meghwal, M., & Goyal, M. R., (eds.), *Food Process Engineering: Emerging Trends in Research and Their Applications* (Vol. 5). As part of book series on "Innovations in Agricultural and Biological Engineering" Apple Academic Press, USA.

Amami, E., Khezami, W., Mezrigui, S., Badwaik, L. S., Bejar, A. K., Perez, C. T., & Kechaou, N., (2017). Effect of ultrasound-assisted osmotic dehydration pretreatment on the convective drying of strawberry. *Ultrasonics Sonochemistry, 36,* 286–300.

Bahnasawy, A. H., & Shenana, M. E., (2004). A mathematical model of direct sun and solar drying of some fermented dairy products (Kishk). *Journal of Food Engineering, 61*(3), 309–319.

Barbosa-Cánovas, G., & Bermúdez-Aguirre, D., (2010). Other milk preservation technologies: Ultrasound, irradiation, microwave, radio frequency, ohmic heating, ultraviolet light and bacteriocins. In: Griffiths, M., (ed.), *Improving the Safety and Quality of Milk* (Vol. 1, pp. 420–450). Woodhead Publishing, Cambridge, UK.

Barman, N., & Badwaik, L. S., (2017). Effect of ultrasound and centrifugal force on carambola (*Averrhoa carambola* L.) slices during osmotic dehydration. *Ultrasonics Sonochemistry, 34*, 37–44.

Botha, G. E., Oliveira, J. C., & Ahrné, L., (2012). Quality optimization of combined osmotic dehydration and microwave assisted air drying of pineapple using constant power emission. *Food and Bioproducts Processing, 90*(2), 171–179.

Cano-Lamadrid, M., Lech, K., Michalska, A., Wasilewska, M., Figiel, A., Wojdyło, A., & Carbonell-Barrachina, Á. A., (2017). Influence of osmotic dehydration pre-treatment and combined drying method on physico-chemical and sensory properties of pomegranate arils, cultivar Mollar de Elche. *Food Chemistry, 232*, 306–315.

Castelló, M. L., Igual, M., Fito, P. J., & Chiralt, A., (2009). Influence of osmotic dehydration on texture, respiration and microbial stability of apple slices (Var. Granny Smith). *Journal of Food Engineering, 91*(1), 1–9.

Chemat, F., Zill, E. H., & Khan, M. K., (2011). Applications of ultrasound in food technology: Processing, preservation and extraction. *Ultrasonics Sonochemistry, 18*(4), 813–835.

Chouliara, I., Savvaidis, I. N., Panagiotakis, N., & Kontominas, M. G., (2004). Preservation of salted, vacuum-packaged, refrigerated sea bream (*Sparus aurata*) fillets by irradiation: Microbiological, chemical and sensory attributes. *Food Microbiology, 21*(3), 351–359.

Ciurzyńska, A., Kowalska, H., Czajkowska, K., & Lenart, A., (2016). Osmotic dehydration in production of sustainable and healthy food. *Trends in Food Science and Technology, 50*, 186–192.

Corrêa, J. L. G., Pereira, L. M., Vieira, G. S., & Hubinger, M. D., (2010). Mass transfer kinetics of pulsed vacuum osmotic dehydration of guavas. *Journal of Food Engineering, 96*(4), 498–504.

Corrêa, J. L. G., Rasia, M. C., Mulet, A., & Cárcel, J. A., (2017). Influence of ultrasound application on both the osmotic pretreatment and subsequent convective drying of pineapple (*Ananas comosus*). *Innovative Food Science and Emerging Technologies, 41*, 284–291.

Da Costa, R. A. S., Aguiar-Oliveira, E., & Maldonado, R. R., (2016). Optimization of osmotic dehydration of pear followed by conventional drying and their sensory quality. *LWT – Food Science and Technology, 72*, 407–415.

Defraeye, T., & Verboven, P., (2017). Convective drying of fruit: Role and impact of moisture transport properties in modeling. *Journal of Food Engineering, 193*, 95–107.

Defraeye, T., (2017). Impact of size and shape of fresh-cut fruit on the drying time and fruit quality. *Journal of Food Engineering, 210*, 35-4.

Dehghannya, J., Naghavi, E. A., & Ghanbarzadeh, B., (2016). Frying of potato strips pretreated by ultrasound-assisted air-drying. *Journal of Food Processing and Preservation, 40*(4), 583–592.

Del Valle, J. M., Aránguiz, V., & León, H., (1998). Effects of blanching and calcium infiltration on PPO activity, texture, microstructure and kinetics of osmotic dehydration of apple tissue. *Food Research International, 31*(8), 557–569.

Deng, Y., & Zhao, Y., (2008a). Effect of pulsed vacuum and ultrasound osmopretreatments on glass transition temperature, texture, microstructure and calcium penetration of dried apples (Fuji). *LWT—Food Science and Technology, 41*(9), 1575–1585.

Deng, Y., & Zhao, Y., (2008b). Effects of pulsed-vacuum and ultrasound on the osmodehydration kinetics and microstructure of apples (Fuji). *Journal of Food Engineering, 85*(1), 84–93.

Dias Da Silva, G., Barros, Z. M. P., De Medeiros, R. A. B., De Carvalho, C. B. O., Rupert, B. S. C., & Azoubel, P. M., (2016). Pretreatments for melon drying implementing ultrasound and vacuum. *LWT – Food Science and Technology, 74*, 114–119.

Duan, X., Zhang, M., Li, X., & Mujumdar, A. S., (2008). Ultrasonically enhanced osmotic pretreatment of sea cucumber prior to microwave freeze drying. *Drying Technology, 26*(4), 420–426.

Erle, U., & Schubert, H., (2001). Combined osmotic and microwave-vacuum dehydration of apples and strawberries. *Journal of Food Engineering, 49*(2&3), 193–199.

Fernandes, F. A. N., & Rodrigues, S., (2008). Application of ultrasound and ultrasound-assisted osmotic dehydration in drying of fruits. *Drying Technology, 26*(12), 1509–1516.

Fernandes, F. A. N., Gallão, M. I., & Rodrigues, S., (2008). Effect of osmotic dehydration and ultrasound pre-treatment on cell structure: Melon dehydration. *LWT – Food Science and Technology, 41*(4), 604–610.

Gallego-Juarez, J. A., Rodriguez-Corral, G., Gálvez Moraleda, J. C., & Yang, T. S., (1999). A new high-intensity ultrasonic technology for food dehydration. *Drying Technology, 17*(3), 597–608.

Garcia-Noguera, J., Oliveira, F. I. P., Gallão, M. I., Weller, C. L., Rodrigues, S., & Fernandes, F. A. N., (2010). Ultrasound-assisted osmotic dehydration of strawberries: Effect of pretreatment time and ultrasonic frequency. *Drying Technology, 28*(2), 294–303.

García-Segovia, P., Mognetti, C., Andrés-Bello, A., & Martínez-Monzó, J., (2010). Osmotic dehydration of Aloe Vera (Aloe barbadensis Miller). *Journal of Food Engineering, 97*(2), 154–160.

Goula, A. M., & Lazarides, H. N., (2012). Modeling of mass and heat transfer during combined processes of osmotic dehydration and freezing (Osmo-Dehydro-Freezing). *Chemical Engineering Science, 82*, 52–61.

Jangam, S. V., (2011). An overview of recent developments and some R&D challenges related to drying of foods. *Drying Technology, 29*(12), 1343–1357.

Jiang, N., Liu, C., Li, D., Zhang, Z., Liu, C., Wang, D., Niu, L., & Zhang, M., (2017). Evaluation of freeze drying combined with microwave vacuum drying for functional okra snacks: Antioxidant properties, sensory quality, and energy consumption. *LWT – Food Science and Technology, 82*, 216–226.

Kachele, R., Zhang, M., Gao, Z., & Adhikari, B., (2017). Effect of vacuum packaging on the shelf-life of silver carp (*Hypophthalmichthys molitrix*) fillets stored at 4°C. *LWT – Food Science and Technology, 80*, 163–168.

Karizaki, V. M., (2016). Kinetic modeling and determination of mass transfer parameters during cooking of rice. *Innovative Food Science and Emerging Technologies, 38*(Part A), 131–138.

Karizaki, V. M., Sahin, S., Sumnu, G., Mosavian, M. T. H., & Luca, A., (2013). Effect of ultrasound-assisted osmotic dehydration as a pretreatment on deep fat frying of potatoes. *Food and Bioprocess Technology, 6*(12), 3554–3563.

Kek, S. P., Chin, N. L., & Yusof, Y. A., (2013). Direct and indirect power ultrasound assisted pre-osmotic treatments in convective drying of guava slices. *Food and Bioproducts Processing, 91*(4), 495–506.

Khir, R., Pan, Z., Salim, A., Hartsough, B. R., & Mohamed, S., (2011). Moisture diffusivity of rough rice under infrared radiation drying. *LWT – Food Science and Technology, 44*(4), 1126–1132.

Kowalski, S. J., & Szadzińska, J., (2014). Convective-intermittent drying of cherries preceded by ultrasonic assisted osmotic dehydration. *Chemical Engineering and Processing: Process Intensification, 82*, 65–70.

Kowalski, S. J., Szadzińska, J., & Pawłowski, A., (2015). Ultrasonic-assisted osmotic dehydration of carrot followed by convective drying with continuous and intermittent heating. *Drying Technology, 33*(13), 1570–1580.

Kung, H. F., Lee, V., Lin, C. W., Huang, Y. R., Cheng, C. A., Lin, C. M., & Tsai, Y. H., (2017). The effect of vacuum packaging on histamine changes of milkfish sticks at various storage temperatures. *Journal of Food and Drug Analysis, 25*(4), 812–818.

Mahato, D. K., Verma, D. K., Billoria, S., Kopari, M., Prabhakar, P. K., Ajesh, K. V., Behera, S. M., & Srivastav, P. P., (2017). Applications of nuclear magnetic resonance in food processing and packaging management. In: Meghwal, M., & Goyal, M. R., (eds.), *Developing Technologies in Food Science Status, Applications, and Challenges* (Vol. 7). As part of book series on *"Innovations in Agricultural and Biological Engineering"* Apple Academic Press, USA.

Mason, T. J., Paniwnyk, L., & Chemat, F., (2003). Ultrasound as a preservation technology. In: Zeuthen, & Bøgh-Sørensen, L., (eds.), *Food Preservation Techniques* (pp. 303–337). As part of book series on food science, technology and nutrition. Woodhead Publishing, Cambridge, UK.

Moreno, J., Chiralt, A., Escriche, I., & Serra, J. A., (2000). Effect of blanching/osmotic dehydration combined methods on quality and stability of minimally processed strawberries. *Food Research International, 33*(7), 609–616.

Mosavian, M. T. H., & Karizaki, V. M., (2012). Determination of mass transfer parameters during deep fat frying of rice crackers. *Rice Science, 19*(1), 64–69.

Mujumdar, A. S., (2004). *Dehydration of Products of Biological Origin*. CRC Press/Taylor & Francis Group, Boca Raton, Florida, USA.

Mujumdar, A., (2014). Principles, classification, and selection of dryers. In: Mujumdar, A. S., (ed.), *Handbook of Industrial Drying* (4th edn., pp. 3–29). CRC Press/Taylor & Francis Group, Boca Raton, Florida, USA.

Musielak, G., Mierzwa, D., & Kroehnke, J., (2016). Food drying enhancement by ultrasound – A review. *Trends in Food Science and Technology, 56*, 126–141.

Nowacka, M., Tylewicz, U., Laghi, L., Dalla, R. M., & Witrowa-Rajchert, D., (2014). Effect of ultrasound treatment on the water state in kiwifruit during osmotic dehydration. *Food Chemistry, 144*, 18–25.

Nowacka, M., Tylewicz, U., Romani, S., Dalla, R. M., & Witrowa-Rajchert, D., (2017). Influence of ultrasound-assisted osmotic dehydration on the main quality parameters of kiwifruit. *Innovative Food Science and Emerging Technologies, 41*, 71–78.

Onwude, D. I., Hashim, N., & Chen, G., (2016). Recent advances of novel thermal combined hot air drying of agricultural crops. *Trends in Food Science and Technology, 57*(Part A), 132–145.

Parniakov, O., Bals, O., Lebovka, N., & Vorobiev, E., (2016). Effects of pulsed electric fields assisted osmotic dehydration on freezing-thawing and texture of apple tissue. *Journal of Food Engineering, 183*, 32–38.

Patel, J. H., & Sutar, P. P., (2016). Acceleration of mass transfer rates in osmotic dehydration of elephant foot yam (*Amorphophallus paeoniifolius*) applying pulsed-microwave-vacuum. *Innovative Food Science and Emerging Technologies, 36*, 201–211.

Pingret, D., Fabiano-Tixier, A. S., & Chemat, F., (2013). Degradation during application of ultrasound in food processing: A review. *Food Control, 31*(2), 593–606.

Porciuncula, B. D. A., Zotarelli, M. F., Carciofi, B. A. M., & Laurindo, J. B., (2013). Determining the effective diffusion coefficient of water in banana (Prata variety) during osmotic dehydration and its use in predictive models. *Journal of Food Engineering, 119*(3), 490–496.

Prosapio, V., & Norton, I., (2017). Influence of osmotic dehydration pre-treatment on oven drying and freeze drying performance. *LWT – Food Science and Technology, 80*, 401–408.

Prothon, F., Ahrné, L. L. M., Funebo, T., Kidman, S., Langton, M., & Sjöholm, I., (2001). Effects of combined osmotic and microwave dehydration of apple on texture, microstructure and rehydration characteristics. *LWT – Food Science and Technology, 34*(2), 95–101.

Rajkumar, G., Shanmugam, S., Galvâo, M. D. S., Dutra, S. R. D., Leite, N. M. T. S., Narain, N., & Mujumdar, A. S., (2017). Comparative evaluation of physical properties and volatiles profile of cabbages subjected to hot air and freeze drying. *LWT – Food Science and Technology, 80*, 501–509.

Ramos, I. N., Miranda, J. M. R., Brandão, T. R. S., & Silva, C. L. M., (2010). Estimation of water diffusivity parameters on grape dynamic drying. *Journal of Food Engineering, 97*(4), 519–525.

Rastogi, N. K., Raghavarao, K. S. M. S., & Niranjan, K., (2014). Recent developments in osmotic dehydration. In: Da-Wen, S., (ed.), *Emerging Technologies for Food Processing* (2nd edn., pp. 181–212). Academic Press, San Diego, USA.

Rastogi, N. K., Suguna, K., Nayak, C. A., & Raghavarao, K. S. M. S., (2006). Combined effect of γ-irradiation and osmotic pretreatment on mass transfer during dehydration. *Journal of Food Engineering, 77*(4), 1059–1063.

Rodrigues, S., Oliveira, F. I. P., Gallão, M. I., & Fernandes, F. A. N., (2009). Effect of immersion time in osmosis and ultrasound on papaya cell structure during dehydration. *Drying Technology, 27*(2), 220–225.

Ruiz-López, I. I., Ruiz-Espinosa, H., Arellanes-Lozada, P., Bárcenas-Pozos, M. E., & García-Alvarado, M. A., (2012). Analytical model for variable moisture diffusivity estimation and drying simulation of shrinkable food products. *Journal of Food Engineering, 108*(3), 427–435.

Seguí, L., Fito, P. J., & Fito, P., (2012). Understanding osmotic dehydration of tissue structured foods by means of a cellular approach. *Journal of Food Engineering, 110*(2), 240–247.

Seth, D., & Sarkar, A., (2004). A lumped parameter model for effective moisture diffusivity in air drying of foods. *Food and Bioproducts Processing, 82*(3), 183–192.

Sette, P., Salvatori, D., & Schebor, C., (2016). Physical and mechanical properties of raspberries subjected to osmotic dehydration and further dehydration by air- and freeze-drying. *Food and Bioproducts Processing, 100*(Part A), 156–171.

Silva, M. A. D. C., Silva, Z. E. D., Mariani, V. C., & Darche, S., (2012). Mass transfer during the osmotic dehydration of West Indian cherry. *LWT – Food Science and Technology, 45*(2), 246–252.

Simal, S., Benedito, J., Sánchez, E. S., & Rosselló, C., (1998). Use of ultrasound to increase mass transport rates during osmotic dehydration. *Journal of Food Engineering, 36*(3), 323–336.

Siucińska, K., & Konopacka, D., (2014). Application of ultrasound to modify and improve dried fruit and vegetable tissue: A review. *Drying Technology, 32*(11), 1360–1368.

Stojanovic, J., & Silva, J. L., (2007). Influence of osmotic concentration, continuous high frequency ultrasound and dehydration on antioxidants, color and chemical properties of rabbit eye blueberries. *Food Chemistry, 101*(3), 898–906.

Tadeusz, K., & Mujumdar, A. S., (2009). Classification and selection criteria. *Advanced Drying Technologies* (2nd edn., pp. 11–17). CRC Press/Taylor & Francis Group, Boca Raton, Florida, USA.

Toğrul, İ. T., & İspir, A., (2008). Equilibrium distribution coefficients during osmotic dehydration of apricot. *Food and Bioproducts Processing, 86*(4), 254–267.

Tontul, I., & Topuz, A., (2017). Spray-drying of fruit and vegetable juices: Effect of drying conditions on the product yield and physical properties. *Trends in Food Science and Technology, 63*, 91–102.

Torreggiani, D., & Bertolo, G., (2004). Present and future in process control and optimization of osmotic dehydration: From unit operation to innovative combined process: An overview. *Advances in Food and Nutrition Research, 48*, 173–238.

Verma, D. K., & Srivastav, P. P., (2017). In: Verma, D. K., & Srivastav, P. P., (eds.), *Microorganisms in Sustainable Agriculture, Food and the Environment* (Vol. 1). As part of book series on "Innovation in Agricultural Microbiology," Apple Academic Press, USA.

Wang, X., Gao, Z., Xiao, H., Wang, Y., & Bai, J., (2013). Enhanced mass transfer of osmotic dehydration and changes in microstructure of pickled salted egg under pulsed pressure. *Journal of Food Engineering, 117*(1), 141–150.

Xin, Y., Zhang, M., & Adhikari, B., (2013). Effect of trehalose and ultrasound-assisted osmotic dehydration on the state of water and glass transition temperature of broccoli (*Brassica oleracea* L. var. botrytis L.). *Journal of Food Engineering, 119*(3), 640–647.

Xu, B., Zhang, M., Bhandari, B., & Cheng, X., (2014). Influence of ultrasound-assisted osmotic dehydration and freezing on the water state, cell structure, and quality of radish (*Raphanus sativus* L.) cylinders. *Drying Technology, 32*(15), 1803–1811.

Yao, Y., (2016). Enhancement of mass transfer by ultrasound: Application to adsorbent regeneration and food drying/dehydration. *Ultrasonics Sonochemistry, 31*, 512–531.

# CHAPTER 8

# Hydrodynamic Cavitation Technology for Food Processing and Preservation

NAVEEN KUMAR MAHANTI,[1] SUBIR KUMAR CHAKRABORTY,[2]
S. SHIVA SHANKAR,[3] and AJAY YADAV[4]

[1]*Agricultural Processing and Structures Division, Indian Agricultural Research Institute, New Delhi, India, E-mail: naveeniitkgp13@gmail.com*

[2]*Agro Produce Processing Division, ICAR-Central Institute of Agricultural Engineering, Bhopal, Madhya Pradesh, India,
E-mail: Subir.Kumar@icar.gov.in*

[3]*Department of Post-Harvest and Food Engineering, G.B. Pant University of Agriculture and Technology, Udham Singh Nagar, Pantnagar, Uttarakhand, India, E-mail: shiva14cae@gmail.com*

[4]*Scientist, Center of Excellence for Soybean Processing and Utilization, ICAR-Central Institute of Agricultural Engineering, Bhopal, Madhya Pradesh, India, E-mail: ajyadav007@gmail.com*

## 8.1 INTRODUCTION

The cavitation refers to the generation and growth of cavities and their subsequent disintegration within an extremely small fraction of time, which emits energy of larger magnitude at the transformation. The reactor faces a very high temperature (1000–10,000 K) and pressure (100–500 bar) during the process of cavitation (Suslick, 1990). Generally, the cavitation can be produced using different techniques like acoustic, hydrodynamic, optic, and particle out of which the hydrodynamic and acoustic technologies are quite popular, whereas the optic and particle technologies are lagging to carry out the required physicochemical modifications in bulk solution. However, the latter two techniques are preferred in single-bubble cavitation (Gogate, 2011). At a laboratory-scale, the ultrasound is widely used cavitation technology; however, it is limited for industrialization (Moholkar and Pandit, 2001b). Hydrodynamic cavitation (HD) is the ideal substitute for ultrasound-based on capital cost, energy requirement and food safety offers numerous advantages in food processing sector (like sterilization and cell disintegration), and water purification treatment (such as disinfection, sludge decomposition and organic molecule oxidation, etc.) (Kalumuck and Chahine, 2000; Balasundaram and Harrison, 2006a, b; Goagate, 2011; Martynenko et al., 2015). The cost and energy requirements of different pasteurization technologies are listed in Table 8.1.

**TABLE 8.1**   Analyzing Several Pasteurization Technologies Based on the Cost, Energy Needs, Production Cost, and Food Safety

| Technology | Cost | Energy Requirement (MJ/m³) | Production Cost ($/m³) | Food Safety Risk |
|---|---|---|---|---|
| Tubular pasteurizer | 50–75000$ | 365–540 MJ/m³ | $10–15 | Low |
| HDC | $15,000 | 300–430 | $8–12 | Low |
| HPP high temperature (95°C) | $1–3 million | 13,350 | $371.65 | Low |
| HPP low temperature (5°C) | $1–3 million | 5000 | $138.07 | Medium |
| PEF | $0.8–1 million | 1000–2500 | $30–70 | Does not work on spores |
| UV | $15,000–30,000 | 700–1000 | $20–30 | Effective only on surface |

(*Source*: Reprinted with permission from Martynenko et al., 2015. © Elsevier.)

This chapter focuses on the fundamentals of hydrodynamic and hydro-thermo-dynamic cavitation and also their working principles. It deals with the procedure of several HD reactors and how the design parameters of orifice and venture and fluid characteristics affect the HD. The scope of this chapter also includes its application in food industries to disintegrate the cells, pasteurize, and sterilize the liquid foods.

## 8.2   MECHANISM IN GENERATION OF CAVITATION

### 8.2.1   HYDRODYNAMIC CAVITATION (HD)

The cavities are generated by passing the liquid through a narrow construction such as in the case of venturi tube or orifice plates under the controlled conditions (Goagate and Pandit, 2001). While decreasing the construction diameter during the liquid flow, the pressure is reduced below the liquid-vapor pressure resulting in the liquid flashes, thus generating the cavities. Moreover, the increased pipe diameter at the downstream side may lead to the raised pressure and reduced liquid velocity, thus resulting to collapse of cavities. The basic principle followed in HD is presented in Figure 8.1.

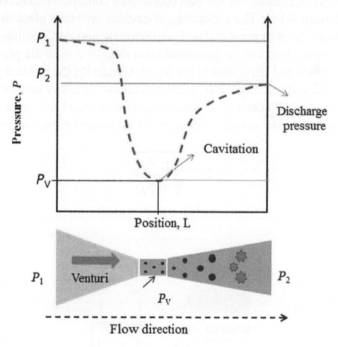

**FIGURE 8.1**   Hydrodynamic cavitation principle.
(*Source*: Reprint from Figure 1 of Carpenter et al. (2016)).

During collapsing, the pressure released is based on inlet pressure, the flow area of orifice/venture and initial radius of nuclei and cavitation intensity is denoted by a dimensionless cavitation number ($C_v$) which is expressed mathematically as:

$$C_v = \frac{p_2 - p_v}{\frac{1}{2}\rho_l v^2{}_0}$$

where, $p_2$ refers to the recovered pressure at the downstream side of orifice, $p_v$ stands for liquid vapor pressure, $v_0$ refers to liquid average velocity at orifice and $\rho_1$ stands for liquid density. The required cavitation intensity based on above-equation can be obtained by regulating the geometric and operating conditions of reactor. The cavitation number at which cavity generation occurs is called a cavitation inception number. If $C_v \leq 1$, the cavitation takes place; however, the best operational condition is achieved when $C_v$ varies from 0.1–1. The generation of cavities will take place in case $C_v$ >2; however, cavities are oscillated continuously and don't collapse. Interestingly, the cavities can be generated even at $C_v = 2$–4 in the presence of dissolved gases and impurities in the liquid. Generally, the reduced pressure and higher liquid velocity are the ideal conditions for cavity generation. The major effects of HD during cavitation are demonstrated in Figure 8.2. The minimum liquid flow velocity, which may lead to cavitation is expressed as:

$$V = \sqrt{\frac{p - p_v}{0.5\rho}}$$

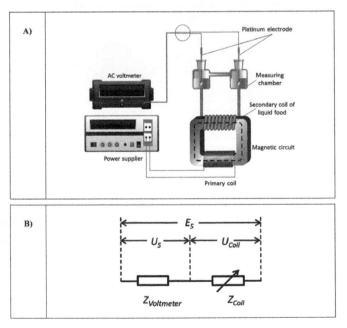

**FIGURE 8.2**  Main effects of hydrodynamic cavitation.
(*Source*: Reprinted from Figure 2 of Cvetkovic et al. (2015)).

### 8.2.2   HYDRO THERMODYNAMIC (HTD) CAVITATION

The mechanical erosion of contact surfaces is the main issue concerned to HD which involves the entrance of high-velocity liquid at cavitation zone, thus inducing the local turbulence resulting in the generation of multi-phase cavitation bubbles (Figure 8.3) and focusing the cavitation at the center of stream. It thus inhibits the erosion of contact metal surfaces and its major advantage is uniform heating throughout the food and thermal enzyme inactivation.

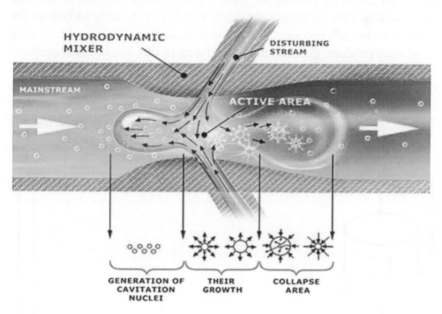

**FIGURE 8.3**   Hydro thermodynamic (HDT) cavitation principle.
(*Source*: Reprinted with permission from Martynenko et al., 2015. © Elsevier.)

## 8.3   EQUIPMENT

### 8.3.1   HIGH PRESSURE HOMOGENIZER (HPH)

In HPH, the disruption of cells occurred due to passing liquid through restricted orifice under high pressure. The High pressure homogenizer (HPH, Figure 8.4) consists of a positive displacement pump, feed tank, and two valves known as 1st stage and 2nd stage where the equipment efficiency is based on the number of passes, operating pressure, suspension temperature and homogenizer valve design (Hetherington et al., 1971; Engler, 1985; Geciova et al., 2002). The liquid is forced by positive displacement pump towards the first-stage valve where the pressure can be raised up to 1000 psi and further increase in pressure is possible bypassing the liquid through second-stage valve where the pressure of nearly 10,000 psi can be achieved. After this, the liquid is again passed to the feed tank so as to recirculate. The increased pressure may result in a higher temperature, which can avoid by passing the cool liquid through the coil in the feed tank. Usually, this technique is widely employed in dairy industries for the purpose of milk homogenization. In addition to cavitation, the flowing liquid may produce the extreme shear force in this kind of homogenizer thus also utilizing the influence of high-velocity liquid over the solid wall (Gogate, 2011). Due to moderate cavitation intensities, the cavitationally active volume and the magnitude of pressure pulses are not properly regulated thus resulting in the generation of cavitation events at the end (Gogate and Pandit, 2001).

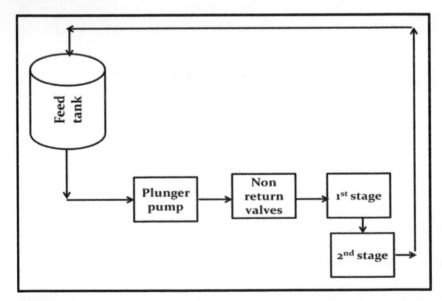

**FIGURE 8.4**   Components and workflow diagram of the high-pressure homogenizer. (*Source*: Reprinted from Gogate and Pandit, (2001)).

### 8.3.2  LOW-PRESSURE HYDRODYNAMIC CAVITATION (HD) REACTOR

Here, the fluid is allowed to pass through a small narrow construction where a sudden decrease in the area results in the increased velocity and decrease in the pressure below the fluid vapor pressure leads to the production of cavities. The low-pressure HD reactor (Figure 8.5) consists of pump control valves which control the pressure, pressure gauges, and construction (either orifice or venture) also. The diameter, number, and shape of holes mainly decide the cavitation intensity and number of cavitational events (Gogate, 2011) and the orifice plate geometry is based on the treatments (disruption or pasteurization).

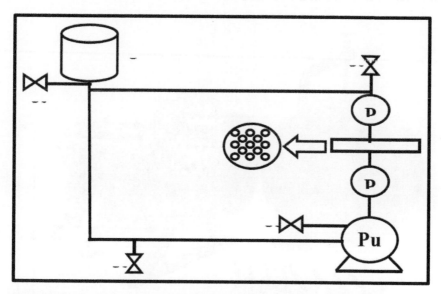

**FIGURE 8.5**    Typical set-up of hydrodynamic cavitation reactor.
(*Source*: Adapted from Lee and Han, (2015)).

### 8.3.3  SHOCK WAVE REACTORS

A shockwave reactor (Rome, Ga., U.S.A) has two cylinders; outer stationary cylinder is separated with annular space from inner rotating cylinder. When inner rotating cylinder rotates at a higher speed while in contact with liquid present in annular space the formation of cavitation takes place. Depressurization of liquid occurs when the rotor forced the liquid into the cavities on the rotor surface. The inner rotating cylinder having multiple cavities causes a sudden change in cross-sectional area leading to depressurization of liquid, which forms the vapors that mixes with the liquid resulting in the formation of bubbles.

Milly et al. (2008) constructed a shockwave reactor (Figure 8.6) to inactivate the microorganisms in apple juice which is operated by a 12 HP motor. In this case, the increase in temperature is totally based on the rotation speed and flow rate of fluid through annular space which is regulated by using frequency speed controller to obtain the desired endpoint. The liquid is fed using a positive displacement pump of 0.25 HP where the pressure of 344.738 kPa was used throughout the experiment.

**FIGURE 8.6**    Schematic representation of the shock wave reactor.
(Source: Reprinted from Milly et al. (2007)).

## 8.4   FACTORS AFFECTING THE SYSTEM

The efficiency and overall cavitation yield of hydrodynamic cavitation are depending on different parameters, and it is shown in Figure 8.7

(Gogate and Pandit, 2001; Dindar, 2016). The effect of different parameters on hydrodynamic cavitation was briefly explained by (Gogate and Pandit, 2001). There are some of geometric parameters affecting hydrodynamic cavitation which are discussed in subsections.

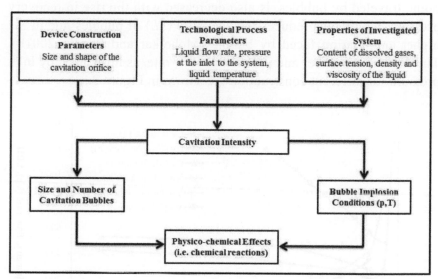

**FIGURE 8.7**    Different factors affecting the intensity of the cavitation process. (*Source*: Reprinted from Ozonek and Lenik, 2011).

## 8.4.1  INLET PRESSURE

The increase in collapse and downstream pressure and energy dissipation rate results in the increase in inlet pressure causing the higher permanent pressure drop across the orifice; therefore, the cavity collapse may produce a higher magnitude of the pressure. On increasing the inlet pressure, there is an increase in pressure at the throat section leading to decreased cavity generation. Due to a reduced number of cavities, the collapse pressure can decrease beyond certain inlet pressure (Gogate and Pandit, 2000), whereas Balasundaram and Harrison (2006a, b) found the decrease in cavitation number with the increasing inlet pressure and cavitation intensity. The enhanced cavitation intensity may cause a sudden rise in a number of cavities (Oba et al., 1986), which leads to the interactions between adjacent cavities, finally decreasing the cavitation efficiency (Guzman et al., 2003). Due to the increased recovery pressure, there is a rise in maximum cavity size before it collapses, thus increasing the cavitation life also (Moholkar and Pandit, 1997). However, the distance from inception point, traveled by bubbles, is also decreased with the rise in recovery pressure, which thus reduces the shelf-life (*SL*) of bubbles due to the higher annihilation faced by bubble (Moholkar and Pandit, 2001b). The cavitation number and volumetric flow rate, as influenced by inlet pressure for different construction are shown in Figure 8.8.

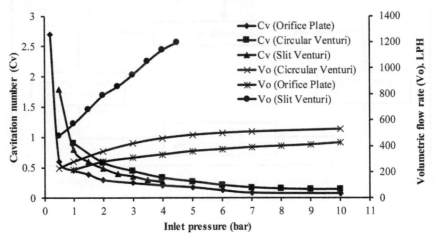

**FIGURE 8.8** Effect of inlet pressure on cavitation number and flow rate for different types of construction.
(*Source*: Reprinted with permission from Saharan et al., 2013. © Elsevier.)

Sahran et al. (2013) observed the decrease in cavitation number on increasing the inlet pressure that also increases the discharge pressure causing the increased flow in the mainline and velocity at the throat section of venturi, which ultimately reduces the $C_v$. The slit venturi has more volumetric flow compared to the orifice plate, and circular venturi for a given pressure drop, whereas the cavitation number is higher in slit and circular venturi than the orifice plate due to more volumetric flows in circular and slit venturi. Several venture-based HD devices, as suggested by Saharan et al. (2013) and Carpenter et al. (2016), are shown in Figure 8.9.

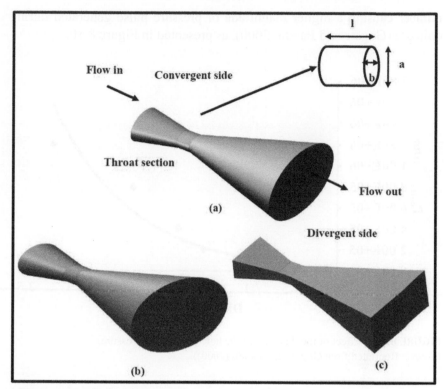

**FIGURE 8.9**   Different hydrodynamic cavitation devices. (a) Elliptical venturi, (b) Circular venturi, and (c) Slit venture.
(*Source*: Reprinted from Carpenter et al. (2016)).

## 8.4.2   EFFECT OF GEOMETRICAL PARAMETERS OF ORIFICE

### 8.4.2.1   DIAMETER OF ORIFICE

Cavitation number increases with increasing diameter of the orifice while the reduction in the flow area may increase the velocity at orifice, thus causing lesser cavitation number but increasing the cavitation intensity and also energy dissipation rate (Balasundaram and Harrison, 2006a, b). There is an increase in collapse pressure as well as cavitation inception number due to the raised orifice diameter (Yan and Thrope, 1990). Larger the diameter of orifice, the higher cavitation number, thus inducing the cavitation. The degree of cavitation also increases for the same cavitation number causing a higher magnitude of pressure pulse generated during collapse (Gogate and Pandit, 2000), as presented in Figure 8.10.

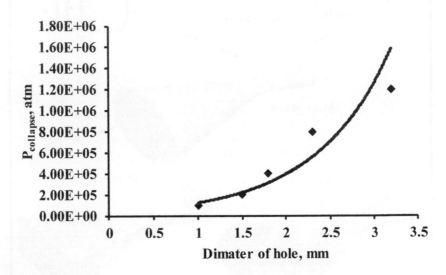

**FIGURE 8.10**    Effect of the diameter of the hole on collapse pressure. (*Source*: Reprinted from Gogate and Pandit (2000)).

### 8.4.2.2    FLOW AREA

The flow area of orifice plate when increased may cause low collapse pressure and reduced velocity at the orifice keeping the main flow rate constant, thus resulting in the slower recovery of downstream pressure and increased cavitation number (Gogate and Pandit, 2000). The increased cavitation number refers to the formation of reduced number of cavities (Senthilkumar et al., 2000; Balasundaram and Harrison, 2006a). The increase in flow area (%) also results in the increased orifice velocity, thus increasing the cavitation intensity and decreasing the cavitation number leading to the raised energy dissipation rate. The increased cavitation enhances the number of cavities, finally resulting in less collapse pressure of cavities. (Oba et al., 1986; Senthilkumar et al., 2000; Balasundaram and Harrison, 2006a).

## 8.4.2.3    SHAPE OF ORIFICE

### 8.4.2.3.1    The Ratio of Perimeter of the Holes to the Total Flow Area (α)

A study by Ozonek and Lenik (2011) has been conducted on the effect of pressure and construction changes on the HD process. In this study, they used different types of orifice plates with different flow areas and geometry (circular and rectangular) with a varying number of holes; it is illustrated in Figure 8.11. The cavitation number usually decreases with an increase in α while on the other side; there is an increase in the cavitation number on increasing the flow area at constant α. The detailed effect of α on the cavitation number is shown in Figure 8.12.

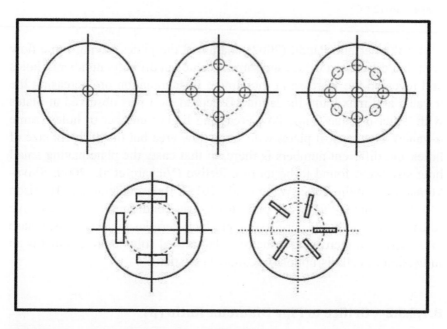

**FIGURE 8.11**    Several kinds of orifice plates employed in the above-investigation. (*Source*: Reprinted from Ozonek and Lenik, 2011).

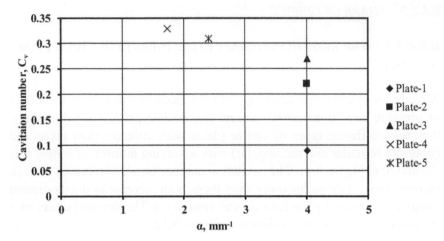

**FIGURE 8.12**    Effect of α on cavitation number at constant operating pressure (7 bar) and temperature (20°C).

Sivakumar and Pandit (2002) examined the plates having same flow areas but different α values and same α value but different number of holes and revealed the increase in rhodamine B degradation on increasing the α value and maximum rhodamine B degradation was observed at plates with larger α value (same flow area) and higher number of holes (same α-value). However, if plates with same flow area but the different size of holes and different numbers is there, in that case, the plate having small hole size were found to be more effective (Vichare et al., 2000; Shivakumar and Pandit, 2002; Wang et al., 2015). Balasubdaram and Harrison (2011) optimized the orifice plate geometric conditions for the selective release of *E. coli* based periplasmic product. Periplasmic acid phosphatase increased with increasing value of α, but a total soluble protein that is an indication of cell disruption was almost constant.

### 8.4.2.3.2    Orifice to Pipe Diameter Ratio (β)

Ozonek and Lenik (2011) reported a rise in cavitation number with the increase in the orifice to pipe diameter ratio; however, there was a decrease in turbulent velocity on increasing the β-value, thus causing enhanced shelf life of bubble (Moholkar and Pandit, 1997). The alteration of the cavitation number with β for several plate geometry is shown in Figure

8.13. Sivakumar and Pandit (2002) suggested the decrease in rhodamine B degradation with the increase in β due to the reduced flow area that enhances the magnitude of collapse pressure. Here, the cavitational yield is also directly proportional to the magnitude of collapse pressure. When two plates have same β value, then plate with smaller hole size produces the more rhodamine B degradation whereas the reduction of hole size increases the turbulent frequency making the collapse of cavities more violent thus increasing the cavitation effect and likewise cavitation yield also (Vichare et al., 2000).

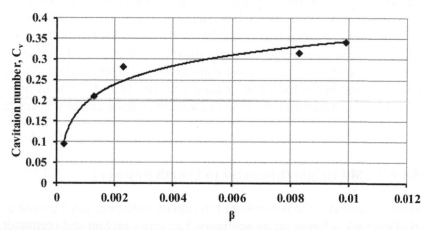

**FIGURE 8.13**    Effect of the orifice to pipe diameter ratio (β) on cavitation number for different geometry of plates at constant operating pressure (7 bar) and temperature (20°C).

The plate offering the highest turbulence or turbulence intensity and shear layer area must be employed in order to improve the cavitation yield (Sivakumar and Pandit, 2002). Moholkar and Pandit (1997) revealed the increase in cavity life before its collapse due to the increase in orifice to pipe diameter ratio (β). When β value = 0.5, there is 73% permanent pressure head loss of orifice pressure which is reduced to 60% for a β value of 0.75. Therefore, the increase in β thus decreases the turbulent intensity making the bubbles less annihilated leading to the increased *SL* of bubble which travels downstream of orifice (Moholkar and Pandit, 2001b).

## 8.4.2.4   EFFECT OF GEOMETRICAL PARAMETERS OF VENTURI

### 8.4.2.4.1   Divergence Angle

In venturi, the divergence angle is the major factor affecting the final collapse pressure of cavities as well as cavitation number and the divergence angle at downstream section can also regulate the pressure recovery rate that is very smooth in venturi compared to orifice due to the design of divergent section at some angle (Jain et al., 2014; Carpenter et al., 2016). In the case of a higher divergence angle, the cavity life reduces resulting in the collapse of cavities quickly causing reduction in cavitation yield due to the recovery of boundary layer separation pressure immediately in the downstream section (Jain et al., 2014; Kuldeep, 2014). On the other hand, the increased divergence angle may lead to increased cavitation number due to the rise in pressure drop at the divergent section that is finally resulting in the reduction of fluid velocity. Thus, the rate of pressure recovery is greater in the case of large divergence angles (Bashir et al., 2011).

### 8.4.2.4.2   Slit Height/Diameter to Length Ratio (γ)

The throat section is the beginning of the cavity initiation, and the generation of cavities is based on its geometry, i.e., cross-section and perimeter to area ratio. The rectangular and elliptical shapes throat have a larger perimeter than the circular throat for the constant cross-sectional area, higher perimeter results in more cavities (Bashir et al., 2011; Kuldeep and Saharan, 2016), so the rectangular and elliptical throats are recommended than the circular throat. The internal area of throat can be expressed using γ- the ratio of throat height/diameter to its length. There is generation of minimum pressure at the throat section. Therefore, the cavities will grow to their maximum resulting in higher magnitude of collapse pressure when there is sufficient internal area of throat. The maximum size attained by cavities and their residence are decided by γ value (Jain et al., 2014; Carpenter et al., 2016). However, the cavity requires higher time in low-pressure region in case of higher length of throat, growing the cavities to a certain size before reaching the downstream section. The length of throat must not exceed the diameter/height of throat for the better operating conditions (Carpenter et al., 2016). The difference between venturi and multihole orifice plate on HD process is listed in Table 8.2.

**TABLE 8.2** Effect of Venturi and Multihole Orifice on Hydrodynamic Cavitation

| Venturi | Multihole Orifice |
| --- | --- |
| Stable cavitation >> Transient cavitation | Transient cavitation >> Stable cavitation |
| Big bubbles | Small bubbles |
| More cavitation events | Lesser cavitation events |
| Efficient pressure recovery | Abrupt pressure recovery. Pressure recovery can be increased by raising the number of holes |
| Mainly mechanical effects | Mechanical and chemical effects (-OH radicals) |
| Relatively lower design flexibility than multihole | Higher design flexibility |

(*Source*: Reprinted from Dahule et al. (2014)).

### 8.4.3 PHYSICOCHEMICAL PROPERTIES OF LIQUID

Numerous parameters of liquid like viscosity, surface tension, cavitation intensity, vapor pressure, temperature, and presence of dissolved gases are affected by the physicochemical properties. The optimum main operating conditions and suitable fluids are selected to increase the cavitational activity. The suggestions for selection of liquid, operating conditions and geometrical factors of cavitation system are given in Table 8.3.

**TABLE 8.3** The Physicochemical Properties of Liquid and Their Favorable Operating Conditions

| Property | Affects | Favorable Conditions |
| --- | --- | --- |
| Surface tension (0.03–0.072 N/m) | Nuclei size | Low surface tension. |
| Viscosity range (1–6 cP) | Transient threshold | Low viscosity. |
| Vapor pressure of liquid (40–100 mm of Hg, 30°C) | Cavitation threshold, cavitation intensity, rate of chemical reaction | Liquid with lower vapor pressures. |

**TABLE 8.3**    *(Continued)*

| Property | Affects | Favorable Conditions |
|---|---|---|
| Bulk liquid temperature (30–70°C) | Intensity of collapse, reaction rate, threshold/ nucleation, physical properties | Optimum value, lower temperatures are preferable. |
| Dissolved gases: Solubility and polytropic constant & thermal conductivity | Gas content, nucleation, collapse phase, Intensity of cavitation events | Lower solubility, Gases having greater polytropic constant and lower thermal conductivity (monoatomic gases). |
| Inlet pressure into the system/rotor speed depending on the type of equipment | — | High rotor speed but operate below the optimum value to avoid super cavitation. |
| Diameter of construction employed in cavity generation | — | Intensive cavitation: higher diameters, reduced intensity: lower diameters having more number of holes. |
| % free area used for flow | — | High cavitation intensity with desired beneficial effects is produced in lower free areas. |

(*Source*: Reprinted from Gogate (2011)).

## 8.5    APPLICATIONS

### 8.5.1    *HYDRODYNAMIC CAVITATION (HD) IN FOOD PROCESSING*

#### 8.5.1.1    *DISRUPTION OF MICROBIAL CELLS*

Microbial cells are sources of enzymes, proteins, and other products. The disruption device that is operated at higher pressure fluid discharge through an orifice is recommended for industrial application. Large scale mechanical cell disruption is high power consuming process, therefore, the HD is more energy efficient technique compared to sonication (Balasundaram and Pandit, 2001). The cavitation effects are of two kinds: physical (shock waves generation, hammer effect, and radial bubble motion) and chemical (free radicals formation) (Balasundaram and Harrison, 2006a, b).

An investigation conducted by Balasundaram and Pandit (2001) compared the energy efficiency in disruption of *S. cerevisiae* cells using several mechanical methods such as sonication, high pressure homogenization, and HD so as to release the invertase. They found the very high (0.12 mg/kJ) specific yield using HPH at 5000 psig for 10 min, whereas the HD had 0.07 mg/kJ specific yield when operated for 10 min that is similar to HPH when operated at 4000 psig. On the contrast, the sonication has very low (0.0019 mg/kJ) specific yield than other two methods. Moreover, the HPH offered higher specific yield of invertase that is 75.9 times of sonication while the specific yield of HD is typically 44 times higher compared to sonication. Again, the sonication treatment is low in specific yield of invertase due to the wall bound nature of invertase and the disruption mechanism of each technique. The invertase yield as affected by disruptive techniques is presented in Table 8.4. The high pressure homogenization derived amount of enzyme (0.0757 mg/ml) was higher than the sonication (0.00531 mg/ml) as well as HD (0.0066 mg/ml) whereas the yield of cytoplasmic proteins was higher found in sonication (0.581 mg/ml) and high pressure homogenization (0.541 mg/ml) than HD (0.1555 mg/ml) because of the broken off only periplasmic wall. Thus, the selective release of invertase can be achieved using the manipulation of operating process parameters in the case of HD.

**TABLE 8.4** Changes in Specific Yield of Invertase Using Different Techniques

| Treatments | Time (min) | Specific Yield (mg/KJ) |
|---|---|---|
| Sonication | 15 | 0.001 |
| | 20 | 0.001 |
| High pressure homogenization | 15 + 3000 psi | 0.06 |
| | 15 + 4000 psi | 0.07 |
| | 10 + 5000 psi | 0.12 |
| Hydrodynamic cavitation | 10 | 0.07 |
| | 20 | 0.04 |
| | 35 | 0.03 |
| | 50 | 0.02 |

(*Sources*: Reprint from Balasundaram and Pandit, (2001)).

Balasundaram and Harrison (2006b) studied the partial disruption of *E. coli* and selective release of specific proteins in addition to the effect of cavitation number, number of passes and growth rate of *E. coli* on the degree and rate of release of acid phosphatise, total soluble protein, and β-galactosidase. Figure 8.14 A demonstrates the effect of cavitation number on the protein and enzymes release from *E. coli*. On the reduction of cavitation number from 0.17–0.13, the release of acid (88%) as well as phosphatase (67%) reaches a constant value. The cavitation intensity if increased, there is generation of more cavities resulting to less collapse pressure, however, the maximum collapse pressure can be achieved at a particular cavitation number. The cavitation efficiency reduces above the critical cavity density due to minimum interaction between the cavities and cells. But the cavitation number between 0.13–0.17 offers a maximum yield of acid phosphatase and β-galactosidase. The release of soluble protein and enzymes from *E. coli* w.r.t., number of passes is illustrated in Figure 8.14B. There is increase in release of protein and β-galactosidase activity up to 600 passes, beyond which there is no effect observed providing the maximum yield of 54% and 59% in case of total soluble protein and β-galactosidase, respectively. The *E. coli* cells when subjected repeatedly to the cavitation zone, the fatigue weakens the cell wall due to the oscillating cavities and their collapse next to cell wall. The constant release rate of soluble proteins was greater by a factor of 2.5 when exposed to the rapidly growing (0.36 h⁻¹) *E. coli* compared to the slowly growing *E. coli* (0.11 h⁻¹). The HD has 4 times lower specific acid phosphatase activity than the osmotic shock at optimum cavitation number 0.17 which refers to the higher release in HD (88%) than osmotic shock (59%). Thus there is more purity of acid phosphate in HD compared to high pressure homogenization and French Press.

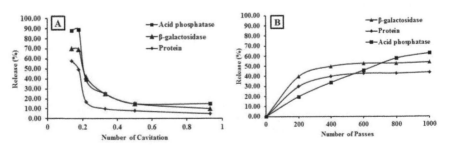

**FIGURE 8.14**   Effect of cavitation number and number of passes on extent release of soluble protein and enzymes from *E. coli*.
(*Source*: Reprinted from Balasundaram and Harrison, (2006b)).

Balasundaram and Harrison (2006a) investigated the effect of process variables on selective release of enzymes form Brewer's yeast at constant cell concentration (1% w/v) over a varying cavitation number (0.09–0.99). The increase in cavitation number results increase in release of soluble protein and enzymes up to cavitation number less than 0.4. The cavitation number when optimized at 0.13 offers the best release of soluble protein and α-glucosidase as well as the maximum cell damage due to generation of maximum collapse pressure. The highest release of total soluble proteins, α-glucosidase and invertase was obtained at 0.5% w/v yeast concentration, 0.13 cavitation numbers and 0.1 to 5% w/v cell concentration.

## 8.5.1.2   PASTEURIZATION AND STERILIZATION OF FOOD

The cellular inactivation was due to induce of physical stress during HD (Milly et al., 2007). The biological entities present in immediate cavitation arca undergo stresses thus causing severe damage to cell walls and finally inactivating the microorganisms (Geciova et al., 2002; Piyasena et al., 2003; Milly et al., 2007). The yeasts are more prone to cavitation effect due to the larger surface area whereas the gram-negative cells are less resistant due to the absence of additional peptidoglycans layer, as found in gram positive cells (Earnshaw, 1998; Milly et al., 2007). On the other side, the spores when compared to vegetative cells are highly resistant to effects of cavitation (Earnshaw, 1998). The HD sterilizes or pasteurizes the liquid food at lower temperatures alone thus offering multiple benefits like efficient energy consumption, microbial inactivation at a lower temperature and quality maintenance (Goagte, 2011).

Milly et al. (2008) observed the inactivation of *Saccharomyces cerevisiae* in apple juice, subjected to sufficient cavitation and found 6.27 mean log cycle reduction in *S. cerevisiae* count at lower temperatures (65.6 and 76.7°C). Energy consumption also decreases to 173 (processing temperature = 65.6°C) and 215 kJ/kg (processing temperature 76.7°C) than the traditional heat treatment (258 kJ/kg). Therefore, this technology can be used for larger scale where the energy saving and efficiency (55–84%) can be enhanced.

Milly et al. (2007) studied the lethality effect of HD on numerous microorganisms at lower temperatures compared to traditional thermal processing techniques for low-acid and high-acid liquid foods including milk and apple and tomato juice. HD induced adequate destructive forces to inactivate vegetative cells of bacteria, yeast, yeast as cospores and heat resistant bacterial spores. The common microorganisms such as yeast, lactic acid bacteria, etc. can be inactivated due to the synergetic effects of cavitation and low temperature. Bacterial spores are more resistant to both thermal and HD; however, the inactivation of spores is possible at lower temperatures using HD. The total lethality obtained in such way is not sufficient for commercial application that can be improved by increasing the inlet fluid temperature.

### 8.5.1.3   OTHER APPLICATIONS

Lohnai et al. (2016) studied the potential of HD reactor to enhance the antioxidant activity (AA) and total phenolic content (TPC) of sorghum flour (SF) and apple pomace (AP). For this, AP, and SF are allowed to ferment naturally followed by HD where authors found that the cavitation temperature and number, flour: water ratio and number of rows of hole in rotor affect the TPC and AA of SF and AP significantly whereas there is a significant effect found in IVSD (*in-vitro* starch digestibility) of SF and TDF of AP, causing no significant impact on TDF (Total dietary fiber). Therefore, the HD technology is considered as the best technology to produce antioxidant rich food materials than the ultrasonic cavitation. Meletharayil et al. (2016) reported the higher water holding capacity of Greek style yogurt (GSY) treated using TMPC (carbon dioxide treated milk protein concentrate) as a protein source along with HD than the commercially strained Greek yogurt. This study exhibits the HD as a suitable method to achieve the comparable titratable acidity values, rheological features and microstructure similar to commercial strained Greek yogurt.

### 8.5.2   HYDRO-THERMODYNAMIC CAVITATION IN FOOD PROCESSING

Recently developed technology – hydro-thermodynamic cavitation is based on the principle of cavitation that is beneficial for crushing, homogenization, and pasteurization of whole food simultaneously. Martynenko et al. (2015) investigated the modifications of chemical characteristics in whole blueberries during crushing, agitation, and heating simultaneously using HD. The HTD processed blueberries contained bioactives similar to fresh blueberry due to the low oxidation. The total phenolics didn't alter even after immediate blueberry processing in HTD at 95°C where the ORAC (oxygen radical absorbance capacity) improved by 15%, anthocyanins dropped by 25% and tannins increased by 65%. HTD included the fine crushing of blueberry along with partial crushing of its seeds. The processed blueberry had viscosity (1.45–2.65 Pa s) preventing the product sedimentation and retained the anthocyanins better than conventionally processed product.

Recently, Martynenko, and Chen (2016) produced the natural value-added puree based products using HTD processing of berries resulting in higher nutritional and physical quality and extended *SL*. The authors reported the higher efficiency of HTD pasteurization than traditional thermal pasteurization and finer texture of HTD blended purees than conventional ones. Moreover, the anthocyanins, and other polyphenolic compounds are found higher in HTD processed blueberry and cranberry puree than their commercial counterparts thus showing very long shelf stability.

## 8.6  SUMMARY AND CONCLUSION

HD, a novel processing technology, is the ideal substitute to ultrasonic-induced cavitation. Theoretically, the optimization of process conditions leads to better results. Numerous reactors can be employed for different purposes; however, the orifice plate with a multi-hole reactor offers huge commercial applications due to its flexibility to achieve the cavitation intensity. The microbial disinfection requires a higher cavitation intensity compared to cell disruption. HD, no doubt, is more energy-efficient than other cavitation techniques, but it causes the erosion of contact parts that can be minimized by focusing the cavitation at the stream center by passing the high velocity liquid at cavitation zone, which is also known as HTD. HTD has huge potential in water processing and limited application in food processing. On contrast, it is a well-established technology at the laboratory and industrial-scale also.

## KEYWORDS

- cavitation
- cavitation intensity
- cavitation number
- cell disruption
- collapse pressure
- food processing
- geometrical parameters
- hydro thermodynamic cavitation
- orifice
- sterilization
- venture

## REFERENCES

Balasundaram, B., & Harrison, S. T. L., (2006a). Disruption of Brewers' yeast by hydrodynamic cavitation: Process variables and their influence on selective release. *Biotechnology and Bioengineering, 94*(2), 303–311.

Balasundaram, B., & Harrison, S. T. L., (2006b). Study of physical and biological factors involved in the disruption of *E. coli* by hydrodynamic cavitation. *Biotechnology Progress, 22*(3), 907–913.

Balasundaram, B., & Harrison, S. T. L., (2011). Optimizing orifice geometry for selective release of periplasmic products during cell disruption by hydrodynamic cavitation. *Biochemical Engineering Journal, 54*, 207–209.

Balasundaram, B., & Pandit, A. B., (2001). Selective release of invertase by hydrodynamic cavitation. *Biochemical Engineering Journal, 8*, 251–256.

Bashir, T. A., Soni, A. G., Mahulkar, A. V., & Pandit, A. B., (2011). The CFD driven optimization of a modified venturi for cavitational activity. *The Canadian Journal of Chemical Engineering, 89*, 1366–1375.

Carpenter, J., Badve, M., Rajoriya, S., George, S., Saharan, V. K., & Pandit, A. B., (2016). Hydrodynamic cavitation: An emerging technology for the intensification of various chemical and physical processes in a chemical process industry. *Reviews in Chemical Engineering.* https://doi.org/10.1515/revce-2016–0032 (Accessed on 8 November 2019).

Cvetkovic, M., Kompare, B., & Klemenčič, A. K., (2015). Application of hydrodynamic cavitation in ballast water treatment. *Environmental Science and Pollution Research, 22*, 7422–7438.

Dahule, R., & Kashinathrao, (2014). *A System and a Process for Water Descaling.* Indian Patent WO/2014/147645.

Dindar, E., (2016). An overview of the application of hydrodynamic cavitation for the intensification of wastewater treatment applications: A review. *Innovative Energy and Research, 5,* 1–7. doi: 10.4172/ier.1000137.

Earnshaw, R. G., (1998). Ultrasound: A new opportunity for food preservation. In: Povey, M. J. W., & Mason, T. J., (eds.), *Ultrasound in Food Processing* (pp. 183–192). Blackie Academic & Professional. London, U.K.

Engler, C. R., (1985). Disruption of microbial cells. In: Moo-Young, M., & Cooney, C. L., (eds.), *Comprehensive Biotechnology* (Vol. 2, pp. 305–324). Pergamon Press, UK.

Geciova, J., Bury, D., & Jelen, P., (2002). Methods for disruption of microbial cells for potential use in the dairy industry: A review. *International Dairy Journal, 12*(6), 541–553.

Gogate, P. R., & Pandit, A. B., (2000). Engineering design methods for cavitation reactors II: Hydrodynamic cavitation reactors. *American Institute of Chemical Engineers Journal, 46*(8), 1641–1649.

Gogate, P. R., & Pandit, A. B., (2001). Hydrodynamic cavitation reactors: A state of the art review. *Reviews in Chemical Engineering, 17,* 1–85.

Gogate, P. R., (2011). Hydrodynamic cavitation for food and water processing. *Food and Bioprocess Technology, 4,* 996–1011.

Guzman, H. R., McNamara, A. J., Nguyen, D. X., & Prausnitz, M. R., (2003). Bioeffects caused by changes in acoustic cavitation bubble density and cell concentration: A unified explanation based on cell-to-bubble ratio and blast radius. *Ultrasound in Medicine and Biology, 8,* 1211–1222.

Hetherington, P. J., Follows, M., Dunnill, P., & Lilly, M. D., (1971). Release of protein from baker's yeast (Saccharomyces cerevisiae) by disruption in an industrial homogenizer. *Transactions of the Institution of Chemical Engineers, 49,* 142–148.

Jain, T., Carpenter, J., & Saharan, V. K., (2014). CFD analysis and optimization of circular and slit venturi for cavitational activity. *Journal of Material Science and Mechanical Engineering, 1*(1), 28–33.

Kalumuck, K. M., & Chahine, G. L., (2000). The use of cavitating jets to oxidize organic compounds in water. *Journal of Fluids Engineering, 122,* 465–470.

Kuldeep, C. J., & Saharan, V. K., (2014). Study of cavity dynamics in a hydrodynamic cavitation reactor. In: Mishra, G. C., (eds.), *Energy Technology and Ecological Concerns: A Contemporary Approach* (pp. 37–43). Gyan Bandhu Publications, New Delhi, India.

Kuldeep., & Saharan, V. K., (2016). Computational study of different venturi and orifice type hydrodynamic cavitating devices. *Journal of Hydrodynamics, 28,* 293–305.

Lee, I., & Han, J. I., (2015). Simultaneous treatment (cell disruption and lipid extraction) of wet microalgae using hydrodynamic cavitation for enhancing the lipid yield. *Bioresource Technology, 186,* 246–251.

Lohani, U. C., Muthukumarappan, K., & Meletharayil, G. H., (2016). Application of hydrodynamic cavitation to improve antioxidant activity in sorghum flour and apple pomace. *Food and Bioproducts Processing, 100,* 335–343.

Martynenko, A., & Chen, Y., (2016). *Hydro Thermodynamic (HTD) Processing of Berries into Natural Foods: Quality and Shelf Life Stability.* CSBE/SCGAB Annual Conference, The Canadian Society for Bioengineering, Paper No. CSBE16–062.

Martynenko, A., Astatkie, T., & Satanina, V., (2015). Novel hydro thermodynamic food processing technology. *Journal of Food Engineering, 152,* 8–16.

Meletharayil, G. H., Metzger, L. E., & Patel, H. A., (2016). Influence of hydrodynamic cavitation on the rheological properties and microstructure of formulated Greek-style yogurts. *Journal of Dairy Science, 99*, 1–12.

Milly, P. J., Toledo, R. T., Kerr, W. L., & Armstead, D., (2007). Inactivation of food spoilage microorganisms by hydrodynamic cavitation to achieve pasteurization and sterilization of fluid foods. *Journal of Food Science, 72*(9), M414–M422.

Milly, P. J., Toledo, R. T., Kerr, W. L., & Armstead, D., (2008). Hydrodynamic cavitation: Characterization of a novel design with energy considerations for the inactivation of *Saccharomyces cerevisiae* in apple juice. *Journal of Food Science, 73*(6), M298–M303.

Moholkar, V. S., & Pandit, A. B., (1997). Bubble behavior in hydrodynamic cavitation: Effect of turbulence. *American Institute of Chemical Engineers Journal, 43*, 1641–1648.

Moholkar, V. S., & Pandit, A. B., (2001a). Modeling of hydrodynamic cavitation reactors: A unified approach. *Chemical Engineering Science, 56*, 6295–6302.

Moholkar, V. S., & Pandit, A. B., (2001b). Numerical investigations in the behavior of one-dimensional bubbly flow in hydrodynamic cavitation. *Chemical Engineering Science, 56*, 1411–1418.

Oba, R., Ikohagi, T., Ito, Y., Miyakura, H., & Sato, K., (1986). Stochastic behavior randomness of desinent cavitation. *Journal of Fluids Engineering, 108*, 438–443.

Ozonek, J., & Lenik, K., (2011). Effect of different design features of the reactor on hydrodynamic cavitation process. *Archives of Material Science and Engineering, 52*(2), 112–117.

Piyasena, P., Mohareb, E., & McKellar, R. C., (2003). Inactivation of microbes using ultrasound: A review. *International Journal of Food Microbiology, 87*(3), 207–216.

Saharan, V. K., Rizwani, M. A., Malani, A. A., & Pandit, A. B., (2013). Effect of geometry of hydro dynamically cavitating device on degradation of orange-G. *Ultrasonics Sonochemistry, 20*, 345–353.

Senthilkumar, P., Sivakumar, M., & Pandit, A. B., (2000). Experimental quantification of chemical effects of hydrodynamic cavitation. *Chemical Engineering Science, 55*(9), 1633–1639.

Sivakumar, M., & Pandit, A. B., (2002). Wastewater treatment: A novel energy efficient hydrodynamic cavitational technique. *Ultrasonics Sonochemistry., 9*, 123–131.

Suslick, K. S., (1990). The chemical effects of ultrasound. *Science, 247*, 1439–1445.

Vichare, N. P., Gogate, P. R., & Pandit, A. B., (2000). Optimization of hydrodynamic cavitation using a model reaction. *Chemical Engineering and Technology, 23*, 683–690.

Wang, Y., Jia, A., Wu, Y., Wu, C., & Chen, L., (2015). Disinfection of bore well water with chlorine dioxide/sodium hypochlorite and hydrodynamic cavitation. *Environmental Technology, 36*(4), 479–486.

Yan, Y., & Thorpe, R. B., (1990). Flow regime transitions due to cavitation in flow through an orifice. *International Journal of Multiphase Flow, 16*(6), 1023–1045.

# CHAPTER 9

# High Pressure Processing (HPP): Fundamental Concepts, Emerging Scope, and Food Application

DEEPAK KUMAR VERMA,[1] MAMTA THAKUR,[2] JAYANT KUMAR,[1]
PREM PRAKASH SRIVASTAV,[1] ASAAD REHMAN SAEED AL-HILPHY,[3]
AMI R. PATEL,[4] and HAFIZ ANSAR RASUL SULERIA[5]

[1]*Agricultural and Food Engineering Department,
Indian Institute of Technology, Kharagpur, West Bengal, India,
E-mails: deepak.verma@agfe.iitkgp.ernet.in; rajadkv@rediffmail.com
(D.K. Verma);
E-mail: Jayant.iitkgpian@gmail.com (J. Kumar);
E-mail: pps@agfe.iitkgp.ernet.in (P.P. Srivastav)*

[2]*Department of Food Engineering and Technology, Sant Longowal
Institute of Engineering and Technology, Longowal, Punjab, India,
E-mails: thakurmamtafoodtech@gmail.com; mamta.ft@gmail.com*

[3]*Department of Food Science, College of Agriculture, University of Basrah,
Basra City, Iraq, E-mail: aalhilphy@yahoo.co.uk*

[4]*Division of Dairy and Food Microbiology, Mansinhbhai Institute of Dairy
and Food Technology-MIDFT, Dudhsagar Dairy Campus, Mehsana, Gujarat,
India, E-mail: amiamipatel@yahoo.co.in*

[5]*UQ Diamantina Institute, Translational Research Institute, Faculty
of Medicine, The University of Queensland, Woolloongabba, Brisbane,
Australia, E-mail: hafiz.suleria@uqconnect.edu.au*

## 9.1 INTRODUCTION

Recently the increased consumer awareness has shifted the food consumption trend from conventional thermal-processed products to the minimally processed foods that are natural, wholesome, microbiologically safe, shelf-stable, and have natural texture. In regard to this, the high pressure processing (HPP) is one such proven technology meeting all such consumer demands and has been gaining popularity in food processing from last few decades. This technique is also known by other names in food industry such as hyperbaric pressure processing, high hydrostatic processing (HHP), ultra-high pressure (UHP) or Pascalization (Linton et al., 2002; Patterson et al., 2006; Ting, 2011; Kadam et al., 2012).

In the times of developing and wielding novel techniques for food processing, HPP is most demanded technology which is nonthermal in nature and demonstrate potential to prolong the shelf-life ($SL$) of food by destroying pathogens, therefore also referred as cold pasteurization (Hoover et al., 1989; Caner et al., 2004; Patterson et al., 2006). The pasteurization of food is possible using HPP when 100–600 MPa pressure is used that leads to the inactivation of microbial cells under the thermodynamic principle of Le Chateliers, equilibrium laws, and isostatic rule, accounting for reduction of microbial load.

The HPP system consists of a central pressure vessel that is a monolithic cylindrical vessel made of high-tensile strength steel and its thickness depends on the lethality of process. The food product after filling and sealing in flexible pouches is kept in the pressure chamber, which utilizes the water as a pressure-transferring medium (Caner et al., 2004; Bermúdez-Aguirre and Barbosa-Cánovas, 2011). The fluid can be pumped extra to create the pressure inside the pressure chamber, and using isostatic principle, the pressure is consistently and uniformly distributed to the entire product in the container ensuring that all parts of food would have equal pressure. Interestingly, the product size and its shape don't have any impact in the HPP treatment.

HPP when compared to conventional heat-based pasteurization process retains the sensory and nutritional characteristics of food similar to their original counterparts (Caner et al., 2004; Palou et al., 2007). However, the proteins and carbohydrates being higher-molecular weight compounds are baro-sensitive but lower-molecular weight compounds like pigments, volatile substances, enzymes, vitamins, etc. are quite resistant to pressure thus resulting in organoleptic and health-promoting effects. The covalent bonds are not affected due to very low energy input in HPP than the thermal treatments.

Moreover, the quality of food is better in HPP compared to other conventionally processed foods thus reduces the use of chemicals as additives significantly for extending the *SL* of food (Hoover et al., 1989; Patterson et al., 2006; Mújica-Paz et al., 2011). The HPP has been marked as a substitute to pasteurization by United States National Advisory Committee (USNAC) on Microbiological Criteria for Foods. The other food safety certification agencies, such as the United States Food and Drug Administration (USFDA) and the United States Department of Agriculture (USDA), have approved the applications of HPP in food (Barbosa-Cánovas and Juliano, 2008). HPP can destroy the pathogens, spoilage causing microbes and enzymes at a comparatively much lower temperature without affecting the other qualities of food, which recommends the use of HPP in the formulation of foods having desired textural and nutritional properties. However, HPP just aids in improving the process efficiency of the conventional thermal methodologies, can't substitute the conventional processing techniques. For instance, the HPP when combined with other techniques such as γ-irradiation, heat, and ultrasound decreases the barosensitivity because of conjugation effects. In addition, HPP can improve the efficiency of processes like dehydration, blanching, rehydration, osmotic dehydration (OD), freezing, and thawing beyond the product novelty and freshness, however, its major limitation is the high initial capital expenditure that may be improved by the reduction of operating costs due to the lesser requirement of energy in HPP. With the advancement of technology, there is gradual reduction in its installation costs. Moreover, the surprising progress in technology transfer may lead to a revolution in foods processing using HPP.

Therefore, this chapter focuses on the fundamental concepts, working principle and equipment used in HPP. The chapter also deals with the application of HPP in food processing and discusses the emerging scope for further research in this field. The present chapter also serves as a reference guide for the individuals working with HPP in research, processing, and new product development in the scientific areas of food technology and engineering.

## 9.2   HIGH PRESSURE PROCESSING (HPP): A BRIEF OVERVIEW

HPP, as stated above, is a processing technique where the food materials is undergone an equal and uniform pressure up to 900 MPa in order to destroy the microbial cells and raising the *SL* of food (Ting, 2011). In this process, the food commodity has been submerged and equally pressurized with the help of water. The high pressure need to apply for particular duration to achieve the desired reduction of microorganisms; for example, 350 and 400 MPa pressure is applied for 30 and 5 minutes, respectively, to reduce the microbial count by 10 times (Hoover et al., 1989). HPP lowers down the microbial load and retain the physic-chemical and sensory properties of food quickly compared to thermal processing (Cano et al., 1997). In the beginning, the HPP was employed mainly in milk preservation; however, later on, fruits and vegetables are also processed using HPP (Palou et al., 2007). High pressure treatment preserves the texture and structure that is why can be used for developing the new products and producing several different types of products have potential to create new products (Rastogi et al., 2007). HPP being nonthermal processing technology prevents the food spoilage, enhances the *SL*, retains the physic-chemical properties (Hoover et al., 1989); however, it is not an ideal technology for low-acid foods including animal and milk-based products (Considine et al., 2008). HPP can be run as a batch or semi-continuous technique in which solid foods are mainly processed using the batch method while the liquid foods are preserved using batch and semi-continuous method (Hogan et al., 2005). Also, HPP is advantageous in the processing of bulky food products compared to conventional thermal methods at the lower temperature based on the different food properties (Balasubramanian and Balasubramaniam, 2003).

### 9.2.1  *WORKING PRINCIPLE AND INSTRUMENTATION OF HPP*

Basically, HPP is the batch operation, and it consists of a pressure vessel that is enclosed on both sides as shown in Figure 9.1. During operation, a yoke mechanism is kept so as to secure the pressure vessel, a pressure intensifier and pump is employed for pressure generation, and a material handling system is used for loading or unloading. Moreover, a special proportional-integral-derivative (PID) system is used to monitor and record the variables to control the process (Ting, 2011).

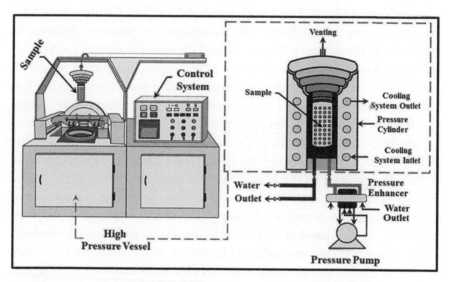

**FIGURE 9.1** Diagrammatic representation of high pressure processing (HPP) equipment. (*Source*: Reprinted with permission from Barba et al., 2016. © Elsevier.)

A retort is mainly used for the operation of HPP where vacuum packaged flexible food pouch is kept inside the cylindrical carrier basket that is placed in the enclosed pressure vessel in which pressure is transmitted on any side of interface. The pressure transmitting medium usually water is pumped to achieve the pressurization using a piston. For this, numerous kinds of pumps, pistons, and intensifiers may be used for achieving the different pressure. In order to obtain the high pressure, the product is kept at a target pressure for particular duration, then quickly vessel is depressurized and the product is unloaded. The cycle time is nearly 10 minutes for the entire process. During compression, the temperature is increased constantly while the temperature is reduced during decompression. This is considered as a distinct advantage in pasteurization and sterilization using HPP (Rasanayagam et al., 2003; Matser et al., 2004; Mújica-Paz et al., 2011).

The water is mainly employed as pressure transmitting fluid during the process and is also compatible with food products. According to Le

Chatlier's principle or "Chatelier's principle" or "The Equilibrium Law," on experiencing a disturbance like concentration, pressure changes, or temperature, the system will automatically respond to restore a new equilibrium state, so that the disturbance effect can be minimized. A system reduces its volume on the application of pressure that makes the covalent bonds to remain unaffected by high pressure even at lower temperature (0–40°C), whereas the hydrophobic and ionic interactions are affected by pressure >200 MPa (Ting, 2011). On the other side, the isostatic principle states the pressure-induced traverses uniformly throughout the directions resulting in the same pressure doses to the products inside the pressure vessel. This theory explains the macroscopic retention of high-moisture non-porous foods using HPP, but the damage is observed in porous foods also due to the variation in air and water compressibility under pressure resulting in alterations of the shape and structure of porous foods such as marshmallows (Balasubramaniam and Farkas, 2008).

## 9.3   APPLICATION OF HPP IN FOOD PROCESSING

### 9.3.1   PROCESSING OF FRUITS, VEGETABLE, AND THEIR PRODUCTS

HPP has emerged as one of the promising technology for the processing of fruits, vegetables, and their products (Table 9.1). Usually, high-acid fruits are the most favorite choices in the HPP operation, which also popularizing the high pressure blanching of fruits and vegetables that minimizes the losses of nutrient quality while leaching. HPP is also highly fruitful in OD that activates the more open food tissue thus improving the mass transfer. The orange juice treated with HPP has a *SL* of 21 days and is microbiologically safe, retaining organoleptic and nutritional properties alike to fresh juice (Oey et al., 2008; Sanchez-Moreno et al., 2009). Moreover, HPP has potential in the pasteurization of fruit juices, vegetable mixes, and RTE meals so as to increase their *SL* of 1–2 months at 4°C (Butz et al., 2002).

**TABLE 9.1** Application of High Pressure Processing in Fruits, Vegetable, and Their Products

| Foods and Food Materials | Optimum Experimental Condition(s) | | | Observations and Concluding Remarks |
| --- | --- | --- | --- | --- |
| | Pressure (MPa) | Holding Time (min) | Temperature (°C) | |
| Fruit and vegetable juices | 400–800 | 10 | 25–50 | No effect of applied heat and pressure over antimutagenicity of strawberry and grapefruit juices; Sensitivity towards heat by antimutagenicity of carrot, leek, spinach, kohlrabi, and cauliflower juices; Excessive high pressure affected the antimutagenicity of beet and tomato juices. |
| Apple juice | 400 | 10 | — | Best sensory properties shown by stored HPP processed apple juice preserved by freezing (−17°C) followed by pressurized compared to pasteurized juice (80°C, 20 min) containing substantial different aroma. |
| Broccoli tomatoes and carrot (crushed or liquid extracts) | 500–800 | — | 25 or 75 | No effect on the Chlorophyll a and b in broccoli, lycopene, and β-carotene in tomatoes as well as antioxidant activities of water-soluble carrot and tomato homogenates; Increase in the water retention and glucose retardation index of tomato pulp; Decrease of carotenoids extractability from coarse carrot homogenates; No significant loss of beneficial substances however there were many structural changes in food matrices. |
| Fresh cut pineapple | 340 | 15 min | — | 3.0, 3.1, and 2.5 decimal reductions of surviving bacteria at 4, 21, and 38°C, respectively; Increasing the shelf-life; <50 cfu/g total plate count, yeast, and mold counts of pressure-treated pineapple cuts. |

**TABLE 9.1** (Continued)

| Foods and Food Materials | Optimum Experimental Condition(s) | | | Observations and Concluding Remarks |
|---|---|---|---|---|
| | Pressure (MPa) | Holding Time (min) | Temperature (°C) | |
| Fruit and vegetable | 100 | 60 | — | Initial texture loss followed by more gradual change due to the instantaneous pulse action of pressure and pressure-hold, respectively resulting in dual effect on fruits and vegetables texture; Pressure treated vegetables were firmer and brighter compared to fresh ones. |
| | 200–500 | 15 | 20–60 | Improved texture of HPP processed and $CaCl_2$ infused carrots by reducing the thermal softening rate. |
| | 400 | 15 | 60 | No texture loss at high temperatures (100–125°C) in HPP processed carrots; Significant reduction in degree of methylation in pressure-treated carrot pectin; Improvement of textural attributes combined with calcium infusion and low-temperature blanching (60°C for 40 min). |
| Guava puree | 600 | 15 | 25 | Refrigerated storage up to 40 days without affecting the color, pectin cloud and ascorbic acid; Retention of water-, oxalate-, and alkali-soluble pectin with original flavor and consistency. |
| Kiwifruit, melon, pears, and peaches | 400 | 30 | 5 or 20 | Best fruit for HPP treatment was melon; Browning was observed in peaches and pears that can be inhibited by ascorbic acid incorporation; Paleness observed in kiwifruit; All fruits have acceptable texture after processing; No inhibition of PPO and POD enzymes by HPP showing higher activities at 20°C than 5°C. |
| Lemon juice | 450 | 2, 5 or 10 | — | No mold growth in HPP treated sample, while the yeast and filamentous fungi spoiled the control sample after 10 days; Minor effects of HPP on composition and physicochemical characteristics. |

**TABLE 9.1** *(Continued)*

| Foods and Food Materials | Optimum Experimental Condition(s) | | | Observations and Concluding Remarks |
|---|---|---|---|---|
| | Pressure (MPa) | Holding Time (min) | Temperature (°C) | |
| Asparagus cauliflower, lettuce, onion, spinach, and tomato | 300–400 | — | — | Reduction of viable aerobic mesophiles, fungi, and yeasts affecting the organoleptic properties; Loosening and peeling of tomato skin keeping the flesh firm without any change in color and flavor; Browning was observed in firm lettuce without altering the flavor; Mild browning observed in cauliflower; Displacement of peroxidase enzyme towards cell interior without inactivation; Better microbial destruction and retention of sensorial characteristics at low temperature and long treatment time. |
| Lychee | 200–600 | 10 or 20 | 20–60 | Minor losses in appearance of fresh and syrup-processed lychee underwent pressure treatment than thermal processing; Increase in POD activity by HPP at 200 MPa beyond which there is no alteration of POD activity; Inactivation of POD and PPO in fresh lychee over 50% and 90%, respectively using pressure of 600 MPa at 60°C for 20 min, but less effects were seen in syrup-processed sample due to baro-protection effect of syrup. |
| Orange Juice | 350 | 1 | 30 | Good quality juice with > 2 months shelf-life at 4°C. |
| | 600 | 1 | 5 | Storage at 0°C up to 20 weeks without affecting the physicochemical and sensory properties; Storage at 10°C induced minor changes after 12 weeks. |
| | 500 | 1 | 5 | Similar microbiological quality (below detectable limits) of HPP treated and thermally pasteurized juice; Better flavor retention when stored at refrigeration for 16 weeks. |

**TABLE 9.1**   *(Continued)*

| Foods and Food Materials | Optimum Experimental Condition(s) | | | Observations and Concluding Remarks |
|---|---|---|---|---|
| | Pressure (MPa) | Holding Time (min) | Temperature (°C) | |
| | 700 | 1 | — | Cloud stabilization in freshly-squeezed orange juice; Attained the shelf-life of 90 days at 4°C. |
| | 400 | 10 | — | Acceptable quality lasted for 150 days storage at ambient conditions. |
| | 500 or 800 | 5 | — | No significant alteration in antioxidant activity, ascorbic acid, sugar, and carotene content when stored upto21 days at 4°C. |
| | 500 | 5 | 35 | Reduced ascorbic acid loss compared to thermally pasteurized (80°C, 30 s) juices. |
| | 350 to 450 | 1–5 | 40 to 60 | Enhanced flavanones extraction and retention of health-promoting properties during cold storage. |
| | 600 | 4 | 40 | Reduced ascorbic acid degradation rate for HPP treated orange juice resulting in better retention of antioxidant activity than conventionally pasteurized juice. |
| | 600 | 1 | 20 | Reduced acceptable levels of aerobic bacteria, yeasts, and other fungi in Navel and Valencia orange juices; 7-log cycle inactivation of Salmonella and marked reduction of PME; No effect on color, browning index, °Brix, viscosity, and titratable acidity, levels of alcohol insoluble acids, ascorbic acid, and β-carotene during storage at 4 or 10°C for 12 weeks. |
| | 600 | 5 | 25 or 80 | Strong protection of folates by excess ascorbates against pressure and heat; Better retention of folates in freshly-squeezed orange juice treated for 5 min at 600 MPa at 25°C. |

**TABLE 9.1** *(Continued)*

| Foods and Food Materials | Optimum Experimental Condition(s) | | | Observations and Concluding Remarks |
|---|---|---|---|---|
| | Pressure (MPa) | Holding Time (min) | Temperature (°C) | |
| Raspberry puree | 200–800 | 15 | 18–22 | Highest anthocyanin stability during pressure treatment at 200 and 800 MPa followed by storage at 4°C. |
| Red and white grape musts | 300–800 | 1–5 | — | Sterilization of white grape must during processing at 500 MPa for 3 min while no complete sterilization of red grape must by HPP treatment at 800 MPa for 5 min; Minor changes observed on account of high-pressure sterilization on physicochemical properties. |
| Strawberry jam | 400 | 5 | Ambient conditions | Mixing, degassing, and pressurization of mixture of powdered sugar, pectin, citric acid and freeze concentrated juice resulted in brighter and red-colored jam that had all original flavor compounds and similar texture to conventional jam.<br><br>Better quality of pressure treated jam and stored at 4°C with minimal losses in sensory and nutritional properties for 3 months than traditional jam. |
| Strawberry juice | 200–500 | — | — | Introduction of new compounds in aroma profile of strawberry during pressure treatment of 800 MPa. |
| | 250–400 | — | — | Significant reduction of strawberry PPO (60%) up to 250 MPa and POD activity (25%) up to 230 MPa during pressurization and depressurization treatments; Optimized destruction of POD was observed at 230 MPa and 43°C. |

**TABLE 9.1**  *(Continued)*

| Foods and Food Materials | Optimum Experimental Condition(s) | | | Observations and Concluding Remarks |
| | Pressure (MPa) | Holding Time (min) | Temperature (°C) | |
| --- | --- | --- | --- | --- |
| Tomato puree | 700 | — | — | Reduction of natural flora below detectable limits; Greater reduction of spore counts of meatballs inoculated with *Bacillus stearothermophilus* spores by HPP than conventional sterilization; No loss of lycopene. |
| | 50–400 | 15 | 25 | High pressure as biggest hurdles along with citric acid and NaCl during manufacturing of minimally processed tomato products having desired sensory and microbiological characteristics; Increased inactivation of PPO, peroxidase, and pectinmethyl esterase with combined treatments at higher pressure values and additive levels. |
| | 100–600 | 12 | 20 | Alteration of lycopene content and presumptive 13-cis isomer (%) by HPP in lycopene solution and tomato puree which was even more in higher temperature storage; Pressurization at 500 MPa and storage at 4 ± 1°C resulted in the highest lycopene stability. |
| White cabbage | 400 or 500 | — | 20, 50 or 80 | Great effect of HPP on soluble and insoluble fiber distribution; Reduced fiber solubility by pressure application up to a temperature of 50°C. |
| White peach | 400 | 10 | 20 | HPP breaks the fruit tissues resulting in enzymic formation of benzaldehyde (due to residual activity of β-glucosidase), C6 aldehydes and alcohols. |

*(Source:* Reprinted with permission from Rastogi et al., 2007. © Taylor & Francis.)

## 9.3.2   PROCESSING OF DAIRY AND DAIRY PRODUCTS

In 1899, Hitefirst revealed the potential of HPP in prolonging the *SL* of milk (Hite, 1899) and found 5–6 log microbial reduction in milk treated at 680 MPa for 10 min under ambient conditions. Numerous investigations have been carried out about the microbial inactivation like *Listeria monocytogenes, L. innocua* and/or *Staphylococcus aureus* in milk (Erkman and Karatas 1997; Gervila et al., 1997) establishing it as a cold pasteurization method. Gram-negative bacteria were less resistant to high pressure either alone or along with nisin compared to other Gram-positive bacteria (Black and Hoover, 2011) mentioning that unlike microbial thermal-resistance, the milk-fat content (0–5%) had no significant effect on microbial lethality. Further, the heat and pressure treatment when combined may result in the development of 'cooked' milk flavor (López-Fandiño, 2006).

There is a huge potential of HPP in milk and dairy-based products including cheese processing (Table 9.2) where it can denature the whey protein, disintegrate the casein micelle, increase the milk pH, reduce the rennet coagulation time (San Martin-Gonzalez et al., 2006). As the milk quality can be improved with HPP and hence it will produce high quality cheese or other end-products as well. Pressure treatment also influences the ripening of few cheese strains. Yogurt from pressure treated milk are creamier than one prepared from conventional methods. Functional molecules are generally demolished by heat that can be preserved using HPP. New functional products made of fruit with antioxidant and similar health specific properties can be launched (Drake et al., 1997; Huppertz et al., 2002).

## 9.3.3   PROCESSING OF MEATS AND MEAT PRODUCTS

The color profile of meat and meat products is particularly affected by HPP because of the denaturation of globin and displacement/release of haem or conversion of ferrous myoglobin to ferric myoglobin on pressurization nearly 400 MPa whereas the pressure treatment up to 400 MPa stabilizes the conformation of egg white and albumin. Thus, HPP in combination with other preservation methods, improves the microbial safety of animal products (Cheftel, 1995; Cheftel and Culioli, 1997; Hugas et al., 2002). HPP also helps in the meat tenderization, coagulation,

**TABLE 9.2**  Application of High Pressure Processing in Dairy and Dairy Products

| Foods and Food Materials | Optimum Experimental Condition | | | Observations and Concluding Remarks |
|---|---|---|---|---|
| | Pressure (MPa) | Holding Time | Temperature (°C) | |
| Cheese | 300–600 | — | — | Increased cheese yield by HPP treatment due to whey proteins denaturation and increased moisture retention; HPP processed cheese had higher moisture content due to disaggregation of casein molecules and fat globules resulting in entrapment of moisture in cheese. |
| | 300–600 | - | - | HPP treated milk led to decreased Cheddar cheese hardness due to the linkage of whey protein with casein in pressurized milk. |
| | 400 | 15 min | 20 | Improved cheese yield (2% d.b.) and moisture content by adjusting pH (7.0) before pressurization. |
| | 200–400 | - | - | Increase in recovery of low-fat cheese from HPP treated milk due to improvement in protein and moisture retention; Improvement of coagulation of pasteurized milk upon pressurization; Quick protein degradation and development of texture and flavor leading to lower hardness and cohesiveness and higher sensory scores. |
| | 400 | 5 min | 21 | Higher interaction levels in HPP treated cheese compared to unpressurized sample; Reduction in L*, a* and b*-values upon HPP after 1 day storage beyond which there is no effect up to 75 days. |
| | 100–800 | 0–60 min | — | No alteration in curd yield up to 250 MPa pressure but 5% decrease in moisture content than untreated milk; Increase in curd yield and moisture content and reduction in whey protein content above 250 MPa pressure. |

**TABLE 9.2** (Continued)

| Foods and Food Materials | Optimum Experimental Condition | | | Observations and Concluding Remarks |
|---|---|---|---|---|
| | Pressure (MPa) | Holding Time | Temperature (°C) | |
| Curd (formation and firming) | 400 | — | — | There was accelerated rate observed in curd formation and firming of rennet milk. |
| | 200–400 | — | — | Enhanced curd firming rate below 200 MPa; Decreased curd firming rate in pressure 200–400 MPa; No effect on gel firmness below 200 MPa that increased at 300 MPa. |
| | 400–600 | — | — | Highest curd formation rate at 200 MPa but minor decrease during pressure range of 400–600 MPa; Enhanced curd firmness observed in the high-pressure treated milk. |
| | 200–500 | 10–110 min | 3–21 | Reduction of water holding capacity and increase in gel strength due to decreased pressure, temperature, and holding time of rennet curd; Highest water holding capacity (40%) and gel strength (0.47 N) was found at 280 MPa, 9°C, and 40 min. |
| | 600 | 30 min | — | Rapid acidification of HPP treated LAB-inoculated pasteurized whole milk compared to unpressurized milk due to faster LAB growth which was due to enhanced amounts of non-sedimentable (non-micellar) caseins, thus increasing the accessible nitrogen supply for bacteria. |
| | 400 | 20 min | 20 | No significant textural and compositional changes in HPP treated Queso fresco cheese except higher moisture content, less firmness, less crumbliness, and more stickiness than the control. |
| | 345–483 | 3 min and 7 min | — | Enhanced shredability of Cheddar cheese by HPP treatment; High pressure produced shreds with better appearance and improved tactile handling. |

**TABLE 9.2** *(Continued)*

| Foods and Food Materials | Optimum Experimental Condition | | | Observations and Concluding Remarks |
|---|---|---|---|---|
| | Pressure (MPa) | Holding Time | Temperature (°C) | |
| Effect on pathogens and spoilage microorganisms | 400–500 | 5–15 min | 2, 10 or 25 | No detection of *E. coli* in the cheese sample from goats' milk inoculated with $10^8$ CFU/g even after 15, 30, or 60 days. |
| | 50–800 | 20 min | 10–30 | Higher sensitivity of *E. coli* in cheddar cheese at pressure above 200 MPa due to acid injury during fermentation; Showing the recovery of sub-lethally injured cells in high-pressure treated cheese slurries. |
| | 400–700 | 1–15 min | — | Reduction of *L. monocytogenes* using HPP treatment in Gorgonzola cheese rinds without affecting the sensory attributes. |
| Microorganisms and enzymes in milk | ~680 | 10 min | Ambient conditions | Prolonged shelf-life due to the 5–6 log cycle reduction in microorganism due to combined effect of pressure and temperature (67–71°C). |
| | 200–250 | — | — | Synergistic effect observed due to the combination of HPP with bacteriocin like lactic in resulted in controlling microbial load of milk without affecting the cheese manufacturing properties. |
| | 200–500 | 60 min | 20 | Effective pathogenic destruction due to periodic pressure oscillation. |
| | 250–450 | 0–80 min | 3 or 21 | Higher microbial destruction in fresh milk due to HPP, longer holding time, and lower temperature resulting in *E. coli* cells least resistant than indigenous microflora. |
| | 250–500 | 5 min | 20 | Higher inactivation of gram-positive bacteria by the combination of high pressure and nisin (0, 250 or 500 iu/ml); Higher sensitivity of gram-negative bacteria to high pressure; Combination of hurdles lower the pressures and shorter treatment times without affecting the product safety. |

TABLE 9.2 *(Continued)*

| Foods and Food Materials | Optimum Experimental Condition | | | Observations and Concluding Remarks |
|---|---|---|---|---|
| | Pressure (MPa) | Holding Time | Temperature (°C) | |
| | 200–1000 | — | — | Increased microbial reduction due to HPP treatment of pasteurized milk (63°C, 30 min) and pasteurization (63°C, 30 min) of HPP treated milk; Higher resistance of milk enzymes to pressure and complete inactivation of alkaline phosphatase and proteinases at 1000 MPa. |
| | 300 | — | — | Reduction of microbes by 4 log-cycles in milk treated with 300 MPa resulting in the shelf-life of 25, 18, and 12 days at 0°C, 5°C, and 10°C, respectively. |
| | 200–600 | 0–120 min | — | Multi-step inactivation of *L. lactis* using HPP and fuzzy logic model to predict the left sublethally damaged cells. |
| | 400 or 500 | 15 min or 3 min, respectively | — | Increased shelf-life of HPP and thermally pasteurized milk by 10 days. |
| | 300–600 | — | — | Minor effect of pressure >300 MPa on β-lactoglobulin; Reduction in whey β-lactoglobulin showing denaturation after 600 MPa. |
| | 400 | 3 min | — | HPP as a gentle process to prolong the milk shelf-life compared to conventional methods without affecting the $B_1$ and $B_6$ vitamins. |
| | 300–500 | — | — | Increased substrate bond availability to plasmin resulted in denaturation of β-lactoglobulin, reduction of plasmin activity and improvement in proteolysis. |

**TABLE 9.2** *(Continued)*

| Foods and Food Materials | Optimum Experimental Condition | | | Observations and Concluding Remarks |
|---|---|---|---|---|
| | Pressure (MPa) | Holding Time | Temperature (°C) | |
| | 300–400 | 0–180 min | — | Enhanced lipoprotein lipase and glutamyl transferase activity on short time pressure exposure; No inactivation of lipase during whole pressure hold time up to 100 mins, whereas $1^{st}$ order inactivation kinetics was followed in glutamyl transferase. |
| | 300–800 | — | 30 to 65 | $1^{st}$ order high-pressure thermal inactivation kinetics of plasmin from milk in two model systems where the first model contained both plasmin and plasminogen while $2^{nd}$ system had all plasminogen converted into plasmin using urokinase; Observation of antagonistic and stabilization effects above 600 MPa due to the disruption of disulfide bonds stabilizing the plasmin and plasminogen structure. |
| | 300–800 | — | 25–65 | $1^{st}$ order kinetics followed in isothermal and high-pressure inactivation of crude plasmin extract of milk at 6.7 pH; Synergistic effect of temperature and high pressure at all temperatures and pressure range from 300–600 MPa; Antagonistic effect between temperature and pressure at pressures > 600 MPa. |
| Rennet coagulation time | 200–400 | — | — | Reduced rennet coagulation time. |
| | 200 | — | — | Reduced rennet coagulation time on pressurization of milk at 200 MPa. |
| | 150–670 | — | — | Constant rennet coagulation time at pressures < 150 MPa beyond which it reduced. |
| | 500–600 | — | — | Greater rennet coagulation time in HPP treated milk compared to pasteurized milk (72°C, 15 s). |

**TABLE 9.2** *(Continued)*

| Foods and Food Materials | Optimum Experimental Condition | | | Observations and Concluding Remarks |
|---|---|---|---|---|
| | Pressure (MPa) | Holding Time | Temperature (°C) | |
| | 250–600 | — | 5 or 10 | Reduced rennet coagulation time by pressure treatment of milk without $KIO_3$ resulting in highest coagulum strength after treatment at 250 or 400 MPa: Highest coagulum strength in milk with $KIO_3$ on pressurization at 400 MPa due to high-pressure induced association of whey proteins with casein micelles. |
| | 100–600 | 0–30 min | 20 | Reduced rennet coagulation time with increasing pressure and time; Higher coagulum strength treated with pressure compared to unheated unpressurized milk; 15% greater yield of cheese curd from pressure treated heated milk than unheated unpressurized milk; 30% less protein content of the whey. |
| Ripening of cheese | 50 | 72 h | — | Increased level of free amino acids due to accelerated cheddar cheese ripening and increased milk protein proteolysis; Excellent taste of pressure treated cheese. |
| | 50 | 72 h | 25 | HPP speeds up the cheddar cheese ripening due to $\alpha_{s1}$–Casein decomposition and $\alpha_{s1}$–1-Casein accumulation. |
| | 200 to 800 | 5 min | 25 | Reduced evolution rate of free amino acids in Cheddar cheeses treated at pressures >400 MPa compared to control; Consistent development of free amino acid at lower pressures; No changes in texture breakdown rate by pressure treatment except treatment at 800 MPa decreased the time-dependent texture changes. |
| | 500 | 15 min | 20 | Higher pH and salt content of cheese prepared from HPP treated goat's milk resulting in rapid maturation and generation of intense flavors; Higher proteolysis observed in high-pressure treated milk-derived cheese due to the formation of small peptides and free amino acids. |

**TABLE 9.2** (Continued)

| Foods and Food Materials | Optimum Experimental Condition | | | Observations and Concluding Remarks |
|---|---|---|---|---|
| | Pressure (MPa) | Holding Time | Temperature (°C) | |
| | 50 or 400 | 72 h or 5 min, respectively | — | Rapid proteolysis and higher pH values of HPP treated goats' milk cheese |
| | 800 | 15 min | 20 | More firm, less fractures and less cohesive cheese from raw and pressure treated goats' milk compared to pasteurized milk (72°C, 15 s); Higher elasticity, constant, and precise protein matrix containing smaller and uniform fat globules were observed in HPP, similar to structure of raw milk-based cheese. |
| | 500 | 15 min | 20 | Gradual increase in the concentration of organic acid (important in cheese flavor) of cheese from HPP treated goat's milk for 2 months. Higher level of organic acids in cheese made from pressure treated or raw milk compared to pasteurized milk. |
| | 50–500 | 20–200 min or 72 h | — | Accelerated Gouda cheese ripening due to HPP treatment. |
| | 50 | 72 h | — | Accelerated free amino acid concentration from HPP treatment, however, pressure of 400 MPa resulted in peptides and casein profiles alike to younger non-treated cheese or cheese treated at 50 MPa; No effect of HPP on plasmin activity but reduction of coagulant activity at 400 MPa was observed. |
| | 200–500 | 10 min | 12 | Destruction of undesirable microorganisms and increased proteolysis in ewe's milk cheese on pressure at 300 MPa; Reduction of inhibition of cheese proteolysis at pressure 400 MPa or more. |

(*Source:* Reprinted with permission from Rastogi et al., 2007. © Taylor & Francis.)

and texturization of fish and meat minces (Balasubramaniam and Farkas, 2008). The higher hardness was exhibited by HPP treated Tiger shrimp samples compared to untreated samples due to myosin molecules denaturation above 100 MPa that accounts for hydrogen and disulfide bonds synthesis in the structures (Rastogi et al., 2007). Moreover, HPP being the single cold pasteurization process doesn't significantly affect the sensory and nutritional composition (Hugas et al., 2002). The major applications of HPP in meat and meat products are presented in Table 9.3.

## 9.4 SUMMARY AND CONCLUSION

A large number of High-pressure processed products are available in the international market (fruit juices, smoothies, milk-shakes, probiotic drinks, and raw squids in Japan, oysters, and mushrooms in the United States, shrimps, and prawns in France). This wide variety and worldwide presence depict the global acceptance of this technology. In the coming future, HPP is going to be commercially used, and its full potential is completely understood. The destruction and inactivation of microbes and enzymes, respectively at low to moderate temperatures without affecting the sensory and nutritional properties exhibit the applications of HPP in the generation of novel and other value-added foods. The severity of HPP process can no doubt be minimized by combination with other techniques like $\gamma$-irradiation, ultrasound, carbon dioxide, anti-microbial peptides, and heat. HPP has also shown the potential in several processing unit operations such as dehydration, blanching, frying, rehydration, gelation, and freezing. However, the major limitation associated with HPP is its higher initial costs, which could be lowered by the operating costs. Consistent modifications are going on the basis of its capacity, throughput, and safety; many industrialists hesitate to invest their money into this technology. However, recent technological success stories on commercial level exhibited a huge potential of HPP in the development of a variety of foods carrying unique texture and flavor. The upcoming years will witness the global adoption of this technology that would benefit consumers based on improved nutritional security, food safety labels, and also processed products at a reasonable cost.

**TABLE 9.3** Application of High Pressure Processing in Meats and Meat Products

| Foods and Food Materials | Optimum Experimental Condition | | | Observations and Concluding Remarks |
|---|---|---|---|---|
| | Pressure (MPa) | Holding Time | Temperature (°C) | |
| Arrowtooth flounder (*Atheresthesstomias*) | 600 | 5 min | — | Most effective stabilizing agent was found to be sorbitol at 8 and 12% against the high-pressure effects on functionality of myofibrillar arrow tooth flounder (*Atheresthesstomias*) proteins compared to sucrose, trehalose, and their mixtures (1:1). |
| Beef | 100–300 | — | — | Modulation of meat proteolytic activities by HPP to improve its quality and enhanced free amino acid content; Increased tryptic digestibility of beef extract at pressure >400 MPa. |
| Beef (post rigor) | 520 | — | — | Enhanced beef toughness on pressurization due to alteration in myofibrillar components against collagen; Reduction of sarcomere length and increased cooking losses on pressurization. |
| Beef (slices) | 100 | 10–15 min | — | Reduction of shear strength, development of pink color, higher score and lowest exudates by HPP treatment. |
| Beef and mutton | 300–700 | 10–20 min | — | Variation in microscopic structure of myofibrils of cattle and mutton muscle and sarcomere shrinkage; Great reduction of shear force values of cattle and mutton skeletal muscle; Improvement in tenderness, natural flavor retention and microbial destruction by making careful choice of temperature, pressure and processing time of HPP. |
| Beef meat | 130–520 | 4.3 min | — | Decrease of total flora and delaying of microbial growth by 1 week on pressurization for a short time resulting in longer maturation and improved meat tenderness; Increased red color improved the meat color by pressure treatment at 130 MPa, i.e., maintained for initial3 days at 4°C without altering the microbiological quality. |

**TABLE 9.3** *(Continued)*

| Foods and Food Materials | Optimum Experimental Condition | | | Observations and Concluding Remarks |
|---|---|---|---|---|
| | Pressure (MPa) | Holding Time | Temperature (°C) | |
| Beef meat (minced) | 200–400 | 20 min | 20 | Reduction of 5 log-cycles or more for *P. fluorescens*, *C. freundii*, and *L. innocua* on pressure treatment >200 MPa, 280 MPa and 400 MPa, respectively; Complete inactivation of *Pseudomonas*, *Lactobacillus*, Coliforms, except the total flora achieved at 400 and 450 MPa leading to reduction by 3–5 log cycles; Delaying of microbial growth for 2–6 days by pressure treatment and storage at 3°C. |
| Beef muscle | 200–800 | — | 20–70 | Increase in hardness with increasing pressure at constant temperature (20–40°C) and temperature at ambient pressure, but hardness decreased at 200 MPa pressure at 60 and 70°C, possibly due to accelerated proteolysis. |
| Blue whiting | 200–420 | 10–30 min | 0–38 | Reduced adhesiveness and paleness and higher water-holding capacity in high-pressure induced blue whiting gels than heat-induced gels; Better elastic gels were produced with the combination of pressure and temperature whereas the harder, more deformed, and more cohesive gels were produced by HPP treatment at chilling temperature. |
| Bovine liver cells | 100–500 | 10 min | 25 | Increased total activities of β-glucuronidase and their acid phosphatase in cytosolic fraction of treated bovine liver cells and of post-rigor treated beef muscles (100–500 MPa, 5 min, 2°C) after pressurization; The lysosomal membranes disruption was used as an index of high pressure effect on catheptic enzymes (affecting meat tenderization) and acid phosphatase; Increased proteinase activity on pressure treatment due to enzyme release from lysosomes. |

**TABLE 9.3**　(Continued)

| Foods and Food Materials | Optimum Experimental Condition | | | Observations and Concluding Remarks |
|---|---|---|---|---|
| | Pressure (MPa) | Holding Time | Temperature (°C) | |
| | 103 | 1–4 min | 30–35 | No effect of pressurization on connective tissue; Meat tenderization on pressure treatment occurred due to improvement of actomyosin toughness; Firmness and contraction achieved on pressurization of ovine and bovine muscles that resulted in more tender meat having high moisture after cooking. |
| Bovine meat | 450 | 15 min | 10 | An intense restructuring effect observed in finely comminuted bovine meats either with or without NaCl subjected to HPP, without any exudation lead to the formation of gels having smooth cohesive texture and high water retention. |
| Chicken and pork batters | 200 and 400 | 30 min | — | Enhanced water and fat-binding features of chicken and pork batters at low-ionic strength; Softer, cohesive, springy, or chewy samples were obtained on pressurization than non-pressurized ones. |
| Chicken breast muscle | 500–800 | 10 min | — | Increased lipid oxidation due to membrane damage on pressure treatment of chicken breast muscle at 800 MPa similar to heat treatment however less oxidation was observed at 600 and 700 MPa; No sign of rancidity up to 500 MPa alike untreated meats recommending 500 MPa as a critical pressure for chicken breast muscle. |
| | 200–600 | 5 | 10 | Hike in secondary lipid oxidation products at 400–600 MPa treatment of cooked breast chicken than 200 MPa and control suggesting the critical influence of storage period in secondary lipid oxidation products generation. |

**TABLE 9.3** *(Continued)*

| Foods and Food Materials | Optimum Experimental Condition | | | Observations and Concluding Remarks |
|---|---|---|---|---|
| | Pressure (MPa) | Holding Time | Temperature (°C) | |
| Chilled cold-smoked salmon | 150–250 | — | — | No destruction of *L. monocytogenes* at 250 MPa but lag phases of 17 and 10 days were achieved at 5 and 10°C, respectively; Significant effect of pressure at 200 MPa on color and texture. |
| | 300 | 20 min | 9 | Reduction of calpains, cathepsin B-like and cathepsin B + L-like enzymes' activities in crude enzyme extracts on pressurization; Achieved complete inactivation of calpain at 300 MPa without affecting proteinase activity; Fish flesh derived enzymes were affected by higher-pressure levels for 18 days and there was an increase in cathepsin B + L-like and calpain activity after 12th day. |
| Cod | 100–400 | — | — | Reduced lipid oxidative stability on pressurization >400 MPa due to metal ions release from complexes; Denaturation of myosin, actin, and most sarcoplasmic proteins was achieved at 100–200 MPa, 300 MPa and 300 MPa, respectively; Survival of many proteinases with reduced activity at 800 MPa; Altered texture (harder, chewier, and gummier) of pressure-treated fish than raw and cooked fish. |
| Cod sausages | 350 | 15 min | 7 | Stable total volatile basic nitrogen during 25 days; No significant pressure effect on microbial counts. |
| Dry-cured Iberian ham and ham slurries | 200–800 | — | — | Initiation of radical generation and lipid peroxidation on low pressure while further reactions and disappearance of free radicals obtained on intermediate to high pressures; Pressure treatment significantly affected the hexanal content like free radicals; Better lightness and reduced redness shown by control samples than pressurized ones. |

**TABLE 9.3** (Continued)

| Foods and Food Materials | Optimum Experimental Condition | | | Observations and Concluding Remarks |
|---|---|---|---|---|
| | Pressure (MPa) | Holding Time | Temperature (°C) | |
| Foie gras (fatty goose or duck liver) | 400 | 10 or 30 min | 50 | Reduced microbial load while maintaining the texture, flavor, and higher product yield; No melting or lipids separation on pressure processing than 15% lipid loss in thermal pasteurization. |
| | 350 and 550 | 1–30 min | 55 and 65 | Equivalent microbiological quality of HPP and temperature to pasteurization on the basis of aerobic mesophilic flora and heat-and pressure-resistant bacterium *Enterococcus faecalis* on duck foie gras. |
| | 100–150 | 1–5 min | 35 | Quick reduction in pH on pressurization soon after slaughter of warm pre-rigor meat. |
| Fresh raw chicken (minced) | 400–900 | 10 min | 14–28 | Reduction of microbial load by 1.7, 3.4, and 3.7 log cycles in fresh raw chicken mince in sealed polyfilm pouches at 408, 616 and 888 MPa, respectively; Chances of microbial spoilage ($10^7$ CFU/g) after 27, 70, and > 98 days, respectively during storage at 4°C. |
| Meat batters with added walnuts | 400 | 10 min | 10 | Enhanced fat content, reduced moisture of meat batters and better water and fat binding attributes on addition of walnuts in cooked meat batters; No effect of HPP on hardness, cohesiveness, springiness, and chewiness but reduced by walnut addition. |
| Meat products (raw sliced marinated beef loin, sliced dry-cured ham, sliced cooked ham) | 600 | 6 min | 31 | Inhibition of Enterobacteriaceae and yeast growth by high pressure reducing the generation of off-flavor in meat products; Also delayed the growth of lactic acid bacteria on pressure treatment; Reduced food safety risks linked with *Salmonella* and *L. monocytogenes*. |

**TABLE 9.3** *(Continued)*

| Foods and Food Materials | Optimum Experimental Condition | | | Observations and Concluding Remarks |
|---|---|---|---|---|
| | Pressure (MPa) | Holding Time | Temperature (°C) | |
| Oysters | 100 and 800 | 10 min | 20 | Destruction of pathogens and increase in pH by HPP in oysters thus reducing the protein and ash contents while increasing the moisture content; After detaching from shells and shucking, the recovered tissue exhibited good shape, more volume, and juiciness compared to untreated oysters. |
| | 400–700 | — | — | Higher baro-resistance of all bacteria associated with illness in oysters than buffer showed no prediction on microbial inactivation in buffer systems. |
| | 350–400 | 1 min | 8.7–10.3 | Inactivation of Hepatitis A virus by 6 log cycle reduction by HPP thus recommending the pressure treatment as a technology for virus reduction. |
| Porcine and bovine | 300–400 | — | — | Improvement in water-holding capacity, increased pH values, and reduction in thermal drip of pressure-treated porcine and bovine. |
| Pork (homogenates) | 400 | 10 min | 25 | Decrease of *E. coli, C. jejuni, P. aeruginosa, S. typhimurium, Y. enterocolitica, S. cerevisiae,* and *C. utilis* populations by 6 log cycles. |
| Pork (paste) | 250 | 20 or 30 min | 20 | Destruction of *Trichinella spiralis* completely at 100 or 150 MPa at 5°C while no inactivation was observed at 50 MPa. |
| Post-rigor beef muscles | 150 | — | 80° | Reduced shear values due to post-rigor beef muscles in a stretched condition because of changes in connective tissue. |
| | Up to 500 | — | — | Easy meat tenderization by HPP without any heating due to enhanced myofibril fragmentation and marked modification in ultrastructure up to 300 MPa, also resulting in the conversion of α-connect into β-connectin and complete degradation of nebulin at 200 MPa. |

**TABLE 9.3** *(Continued)*

| Foods and Food Materials | Optimum Experimental Condition | | | Observations and Concluding Remarks |
|---|---|---|---|---|
| | Pressure (MPa) | Holding Time | Temperature (°C) | |
| Prawns | 200–400 | — | — | Increase in shelf-life up to 28 and 35 days on pressurization at 200 and 400 MPa, respectively than 7 days for air-stored samples whereas extension of shelf-life up to 21 days in vacuum-packaged samples. |
| Pre-or post-rigor | 103.5 | 2 min | 35 | A length reduction of 35–50%, i.e., intense contraction muscles and severe disruption in meat structures on pressure treatment in early pre-rigor stage; Severe modifications in the sarcomere structure without any contraction on pressurization in the post-rigor stage. |
| Rabbit muscles (pre-rigor) | 200–400 | 5 min | — | Reduction of calpains activity on enzymatic extract treatment from pre-rigor rabbit muscles; Higher sensitivity to pressure shown by calpastatin—a specific calpain inhibitor accounting to the maintenance of calpain activity to a certain degree for pressurization < 200 MPa. |
| Raw ham | 400–600 | 5 min | — | Greater shear resistance in control samples than pressurized samples with increasing storage time; No change in the sensory shelf-life of sliced cooked ham processed at 600 MPa for 5 min even after stored for 30 days at 3 or 9°C. |
| Ready-to-eat meats (low-fat pastrami, Strassburg beef, export sausage, and Cajun beef) | 600 | 3 min | 20 | Increase the shelf-life of RTE meats by high pressure during refrigerated storage; Reduction of *L. monocytogenes* by > 4 log cfu/g in inoculated products. |

**TABLE 9.3** *(Continued)*

| Foods and Food Materials | Optimum Experimental Condition | | | Observations and Concluding Remarks |
|---|---|---|---|---|
| | Pressure (MPa) | Holding Time | Temperature (°C) | |
| Salmon mince | - | — | — | Reduction of *L. innocua* by 2.5 log cycle in frozen samples at −28°C for 24 h followed by pressurization and depressurization; A reduction of 1.2 log cycle achieved on pressure-assisted thawing at 207 MPa at 10°C for 23 min; No induction of sub-lethal injury of *L. innocua* by combined high pressure-sub-zero temperature treatments of smoked salmon mince. |
| Salmon spread | 700 | 3 min | — | Prolonged shelf-life from 60–180 days at 3 or 8°C on pressure treatment of salmon spread without affecting the chemical, microbiological, and sensory properties significantly; Complete destruction of pathogens by HPP. |
| Sausages | 500 | 5 or 15 min | 65 | HPP treated sausages were less firm, more cohesive, had lower weight loss, and higher preference scores than heat-treated (40 min at 80–85°C) sample, without compromising the color properties. |
| Sliced Parma ham | 600 | 3, 6 or 9 min | — | Proper control of *L. monocytogenes* by HPP that exhibited the less red color and more intense salty taste. |
| Squid (*Todaropsiseblanae*) | 300 | 20 min | 7 | Increased proteolytic activity and no modification of optimal pH (pH 3) and temperatures on pressure treatment; Pressurization mainly affected the acid cysteine and acid serine proteases, degrading myosin at all temperatures and actin was susceptible to proteolysis only in pressure-treated muscle at 7°C and 40°C. |

**TABLE 9.3** *(Continued)*

| Foods and Food Materials | Optimum Experimental Condition | | | Observations and Concluding Remarks |
|---|---|---|---|---|
| | Pressure (MPa) | Holding Time | Temperature (°C) | |
| Meat tenderization | 100–300 | 10 min | Room temperature | Reduction of μ-calpain level during aging; Reduction of total calpain activity without inducing significant changes in acid phosphatase and alkaline phosphatase activities in pressurized samples compared to control. |
| | 100–200 | — | — | Rapid inactivation of calpastatin at 100 MPa than calpains that remained in muscle treated up to 200 MPa, whereas there was a decrease in calpastatin levels by the pressure treatment. Calpains resisted the pressurization at 200 MPa, however, inactivated at pressure >200 MPa; Increase of total calpains activity in pressurized muscle by HPP resulting in meat tenderization. |
| Tilapia fillets | 50–300 | 12 h | — | Better freshness exhibited by Tilapia fillets stored at 200 MPa for 12 h than control; Also reduction of fillets' total plate count from 4.7–2.0 log cfu/g stored at 200 MPa. |
| | 25–150 | 5 min | 0 | Increased the water-holding capacity of meat homogenates in NaCl solution and increased binding between the meat particles in patties after cooking due to the pressure-induced disaggregation and proteins unfolding. |
| Turkey thigh muscle | Up to 500 | 10 and 30 min | 10 | Exponential dependency of formation of TBA reactive substances with pressure treatment at 10 and 30 min in turkey thigh muscle during storage at 5°C. Apparent activation volume is supposed to quantify the pressure effects on lipid oxidation in meat during further storage. |

*(Source:* Reprinted with permission from Rastogi et al., 2007. © Taylor & Francis.)

## KEYWORDS

- **dairy products**
- **food spoilage**
- **high pressure processing**
- **meat products**
- **shelf-life**

## REFERENCES

Balasubramaniam, V. M., & Farkas, D., (2008). High-pressure food processing. *Journal of Agarochemistry and Food Technology, 14*(5), 413–418.

Balasubramanian, S., & Balasubramaniam, V. M., (2003). Compression heating influence of pressure transmitting fluids on bacteria inactivation during high pressure processing. *Food Research International, 36,* 661–668.

Barba, F. J., Zhu, Z., Koubaa, M., & Sant, A. A. S., (2016). Green alternative methods for the extraction of antioxidant bioactive compounds from winery wastes and by-products: A review. *Trends in Food Science and Technology, 49,* 96–109.

Barbosa-Cánovas, G. V., & Juliano, P., (2008). Food sterilization by combining high pressure and thermal energy. In: Gutiérrez-López, G. F., Barbosa-Cánovas, G. V., Welti-Chanes, J., & Parada-Arias, E., (eds.), *Food Engineering: Integrated Approaches* (pp. 9–46). Springer, New York, USA.

Bermúdez-Aguirre, D., & Barbosa-Cánovas, G. V., (2011). An update on high hydrostatic pressure, from the laboratory to industrial applications. *Food Engineering Reviews, 3*(1), 44–61.

Black, E. P., & Hoover, D. G., (2011). Microbiological aspects of high pressure food processing. In: Zhang, H. Q., Barbosa-Canovas, G. V., Balasubramaniam, V. M., Dunne, C. P., Farkas, D. F., & Yuan, J. T. C., (eds.), *Nonthermal Processing Technologies for Food* (pp. 51–71). Wiley-Blackwell, Oxford.

Butz, P., Edenharder, R., Garcia, A. F., Fister, H., Merkel, C., & Tauscher, B., (2002). Changes in functional properties of vegetables induced by high pressure treatment. *Food Research International, 35,* 295–300.

Caner, C., Hernandez, R. J., & Harte, B. R., (2004). High-pressure processing effects on the mechanical, barrier and mass transfer properties of food packaging flexible structures: A critical review. *Packaging Technology and Science, 17,* 23–29.

Cano, M. P., Hernandez, A., & De Ancos, B., (1997). High pressure and temperature effects on enzyme inactivation in strawberry and orange products. *Journal of Food Science, 62,* 85–88.

Cheftel, J. C., & Culioli, J., (1997). Effects of high pressure on meat: A review. *Meat Science, 46*(3), 211–236.

Cheftel, J. C., (1995). Review: High-pressure, microbial inactivation and food preservation. *Food Science and Technology International, 1,* 75–90.

Considine, K. M., Kelly, A. L., Fitzgerald, G. F., Hill, C., & Sleator, R. D., (2008). High-pressure processing-effects on microbial food safety and food quality. *FEMS Microbial. Lett., 281*(1), 1–9.

Drake, M. A., Harrison, S. L., Asplund, M., Barbosa-Canovas, G., & Swanson, B. G., (1997). High pressure treatment of milk and effects on microbiological and sensory quality of Cheddar cheese. *Journal of Food Science, 62*(4), 843–860.

Erkman, O., & Karatas, S., (1997). Effect of high hydrostatic pressure on staphylococcus aureus in milk. *Journal of Food Engineering, 33,* 257–262.

Gervila, R., Capellas, M., Ferragur, V., & Guamis, B., (1997). Effect of high hydrostatic pressure on *Listeria innocua* 910 CECT inoculated into ewe's milk. *Journal of Food Protection, 60,* 33–37.

Hite, B. H., (1899). The effect of pressure in the preservation of milk. *Washington, Va. University, Agriculture Experiment Station, Bulletin, 58,* 15–35.

Hogan, E., Kelly, A. L., & Sun, D. W., (2005). High pressure processing of foods: An overview. In: Sun, D. W. (ed.), *Emerging Technologies for Food Processing* (pp. 3–31). Academic Press, USA.

Hoover, D. G., Merick, C., Papineau, A. M., Farkas, D. F., & Knorr, D., (1989). Application of high hydrostatic pressure on foods to inactivate pathogenic and spoilage organisms for extension of shelf life. *Food Technology, 43*(3), 99.

Hugas, M., Garriga, M., & Monfort, J. M., (2002). New mild technologies in meat processing: High pressure as a model technology. *Meat Science, 62*(3), 359–371.

Huppertz, T., Kelly, A. L., & Fox, P. F., (2002). Effects of high pressure on constituents and properties of milk. *International Dairy Journal, 12*(7), 561–572.

Kadam, P. S., Jadhav, B. A., Salve, R. V., & Machewad, G. M., (2012). Review on the high pressure technology (HPT) for food preservation. *Journal of Food Processing and Technology, 3*(1), 1–5.

Linton, M., Patterson, M. F., & Patterson, M. F., (2000). High pressure processing of foods for microbiological safety and quality. *ActaMicrobiologicaet Immunologica Hungarica, 47*(2&3), 175–182.

López-Fandiño, R., (2006). High pressure-induced changes in milk proteins and possible applications in dairy technology. *International Dairy Journal, 16*(10), 1119–1131.

Matser, A. M., Krebbers, B., Van Den Berg, R. W., & Bartels, P. V., (2004). Advantages of high-pressure sterilization on quality of food products. *Trends in Food Science and Technology, 15,* 79–85.

Mújica-Paz, H., Valdez-Fragoso, A., Samson, C. T., Welti-Chanes, J., & Torres, J. A., (2011). High-pressure processing technologies for the pasteurization and sterilization of foods. *Food and Bioprocess Technology, 4*(6), 969–985.

Oey, I., Lille, M., Van Loey, A., & Hendrickx, M., (2008). Effect of high-pressure processing on color, texture and flavor of fruit-and vegetable-based food products: A review. *Trends in Food Science and Technology, 19*(6), 320–328.

Palou, E., Lopez-Malo, A., Barbosa-Canovas, G. V., & Swanson, B. G., (2007). High-pressure treatment in food preservation. In: Rahman, M. S., (ed.), *Handbook of Food Preservation* (pp. 815–853). CRC Press, New York, USA.

Patterson, M. F., Ledward, D. A., & Rogers, N., (2006). High-pressure processing. In: Brennan, J. G., (ed.), *Food Processing Handbook* (pp. 173–197). Wiley-Vch, Wokingham.

Rastogi, N. K., Raghavarao, K. S. M. S., Balasubramaniam, V. M., Niranjan, K., & Knorr, D., (2007). Opportunities and challenges in high pressure processing of foods. *Critical Reviews in Food Science and Nutrition, 47*(1), 69–112.

San Martin-Gonzalez, M. F., Welti-Chanes, J., & Barbosa-Canovas, G. V., (2006). Cheese manufacture assisted by high pressure. *Food Reviews International, 22*(3), 275–289.

Sanchez-Moreno, C., De Ancos, B., Plaza, L., Elez-Martinez, P., & Cano, M. P., (2009). Nutritional approaches and health-related properties of plant foods processed by high pressure and pulsed electric fields. *Critical Reviews in Food Science and Nutrition, 49*(6), 552–576.

Ting, E., (2011). High pressure processing equipment fundamentals. In: Zhang, H. Q., Barbosa-Cánovas, G. V., Balasubramaniam, B., Dunne, C. P., Farkas, D. F., & Yuan, J. T., (eds.), *Nonthermal Processing Technologies for Food* (pp. 20–27). Wiley-Blackwell, Oxford.

# CHAPTER 10

# Induced Electric Field (IEF) as an Emerging Nonthermal Techniques for the Food Processing Industries: Fundamental Concepts and Application

NA YANG and DAN-DAN LI

*School of Food Science and Technology, Jiangnan University, Wuxi, Jiangsu Province, China, E-mail: yangna@jiangnan.edu.cn (N. Yang); E-mail: lidandanthora@gmail.com (D.D. Li)*

## 10.1 INTRODUCTION

Nowadays, consumers' desires for nutritious and safe food are motivating engineers and manufacturers to develop new process technologies. Avoidance of rigorous heating during the treatment of food could be achieved via nonthermal methods (Martín-Belloso et al., 2014). Among these technologies, the process of direct application of electric fields (pulsed electric field) to the food has proven to be an effective preservation process capable of providing relatively high-quality products, achieving the maximum retention of their nutritional and sensorial properties (Vega-Mercado et al., 1997). Method for treating foods with electric fields has resulted in numerous applications for dehydration, fermentation, extraction, modification, and sterilization (Timmermans et al., 2014; Dalvi-Isfahan et al., 2016; Dermesonlouoglou et al., 2016; El Darra et al., 2016; Mahato et al., 2017).

Current electrotechnologies for food process are based on electric-potential difference induced by metal electrodes during AC and DC treatment; in particular, these applications include ohmic heating (OH) (Goullieux and Pain, 2005) and moderate electric field (MEF) (Samaranayake and Sastry, 2016a) and high-intensity pulsed electric field

(HIPEF) (Caminiti et al., 2012). OH and HIPEF applications are differentiated by device configurations well as by the thermal or nonthermal effect on processing medium. Electricpotential difference between parallel-plate electrodes excited by sinusoidal or DC voltage results in a thermal effect (OH, typically lower than 1 kV/cm) during the treatment (Jaeger et al., 2016), whereas the non-uniformity distribution of electric field between needle electrodes excited by pulsed voltage causes a nonthermal effect (PEF, typically several kV/cm) (Elez-Martínez et al., 2012). However, *the Joule* heating effect can also be recorded on the edge of the electrode during HIPEF treatment (Ohshima et al., 2016). Electrodes with different material, size, thickness, and arrangement cause various electric field distributions, which induce a differential electrical effect on the foods.

As the food located in a pair of electrodes connected to a power supply during the process, electrochemical reaction between electrode and liquid sample happen immediately (Stancl and Zitny, 2010). A micro-layer is formed on the surface of metal electrode due to oxidation-reduction reactions. It is possible that metal ions pollute the food through electrochemical reactions with high current density. But, electrode corrosion and metal ion leakage can be reduced either by using gold or platinum electrodes as well as by using an increased frequency higher than the power frequency of 50 and 60 Hz (Pataro et al., 2014).

Another method for producing varying voltage is the application of alternating magnetic flux, which has many applications in devices, including inductors, transformers, electric motors, and generators (Tomczuk and Koteras, 2011; Yu et al., 2017). A transformer can transfer electrical energy between the primary and secondary side through magnetic flux without any frequency change. A varying voltage applied to the primary coil induces a magnetic flux in the magnetic circuit. Thereby, an induced voltage (electromotive force, EMF) is generated in the secondary coil (Suresh and Panda, 2016). If a load is connected to a secondary coil that permits the flow of an induced current, this current creates "winding loss" and "load loss" in the secondary coil and the load, respectively. The terms are applied regardless of whether the coil and load are made of copper or another conductive medium, such as food and electrolyte solution. These losses are caused by *Joule* heating and represent an undesirable energy transfer in the energy industry (Villén et al., 2013; Khabdullin et al., 2017). Through rational system design, induced electric fields (IEFs) (potential difference) induced by various magnetic fluxes in liquid sample could possibly be an alternative approach for nonthermal processing.

## 10.2  INDUCED ELECTRIC FIELD (IEF) PROCESSING THEORY

### 10.2.1  INDUCTIVE METHODOLOGY

An alternating electric field is induced by alternating magnetic field as per the Faraday's law of electromagnetic induction (Saslow, 2002). Based on this characteristic, the potential difference between two coils is controllable by differential magnetic fluxes. For the treatment of foods, magneto- IEF could be utilized via the experimental transformer system. The production of IEFs in liquid foods with various field strengths and frequencies is feasible through synchronous alternating magnetic fluxes. A transformer contains a primary coil and a secondary coil, linked by a magnetic circuit. According to its working principle, induced voltages are generated in the two coils because of the changing flux. Figure 10.1 shows a typical transformer containing primary coil, secondary coil, and magnetic circuit (or the magnetic core). An alternating voltage is applied to the primary coil, which sets up an alternating flux (magnetomotive force, MMF) (Heathcote, 2007).

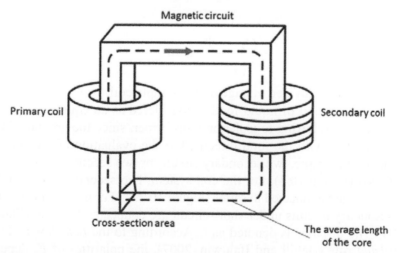

**FIGURE 10.1**    Schematic diagram of single-phase transformer.

## 10.2.2   POTENTIAL DIFFERENCE

Electrical parameters are valuable indices for evaluating food processing under electric field (Fryer, 1995; Marra et al., 2009; Chandrasekaran et al., 2013; Watanabe et al., 2017). Some attempts at electrical measurements for food detection introduce sophisticated instruments (Kuson and Terdwongworakul, 2013; Nakonieczna et al., 2016; Regier et al., 2017). Investigations into the IEF influenced by excitation voltage, frequency, coil configuration, winding direction of the coil, and temperature facilitate the quantitative evaluation of the potential difference in liquid sample. The output voltage between secondary coils of the food is a key technical indicator for IEF processing.

Different from conventional electroanalysis, the magnetic flux is used as a stimulus during the measurement (Yang et al., 2015). In the single-phase system, the output voltage of the secondary coil of the liquid sample can be measured by a voltmeter equipped with electrodes (Figure 10.2A). According to *Ohmic* law, the $E_S$ will be divided on the coil ($Z_{Coil}$) and the $Z_{Voltmeter}$ (Figure 10.2B). The measured voltage ($U_S$, output voltage) developed across the load resistor (the voltmeter) changes due to the physicochemical properties of the sample vary. Their relationship is expressed by the following equations:

$$E_S = U_{Coil} + U_{Voltmeter} \tag{1}$$

$$\frac{U_S}{U_{Coil}} = \frac{Z_{Voltmeter}}{Z_{Coil}} \tag{2}$$

where, $U_{Coil}$ is the divided voltage on the sample coil.

A secondary coil of liquid food is considered as an equipotential cell under the flux, referring to isolation transformer, since the ground which has not contact with the food directly. For the evaluation of the voltage ($U_{ab}$) in liquid sample, two secondary circuits or two secondary coils could be utilized (Figure 10.3). For the coil system, two secondary coils under open-circuit are connected in series (Figure 10.3A). For the circuit system, two secondary circuits under short-circuit are connected together (Figure 10.3B). The distance is denoted as $L$. According to the *Lenz* law and the *right-hand* rule (Bahill and Baldwin, 2007), the polarities of $E_S$ depend upon the direction of the sample coil (or the secondary coil).

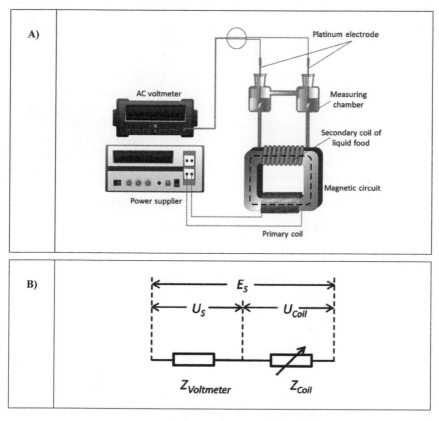

**FIGURE 10.2**  Measurement under an alternating flux. (A) Single-phase system for output voltage measurement, and (B) Equivalent circuit of the measurement.

Under synchronous flux, different phase differences (0° and 180°) between induced voltages ($E_{S1}$ and $E_{S2}$) is obtained through specific coil configuration—namely, in-phase configuration (Figure 10.4A) and reverse-phase configuration (Figure 10.4B). For the in-phase configuration (0°, phase-difference), the two connected coils have the same winding direction (anti-clockwise plus anti-clockwise). For the reverse-phase configuration (180°, phase difference), the two connected coils have different winding directions (clockwise plus anti-clockwise). Their equivalent circuits are shown in Figure 10.4C.

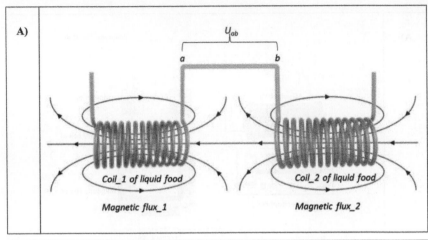

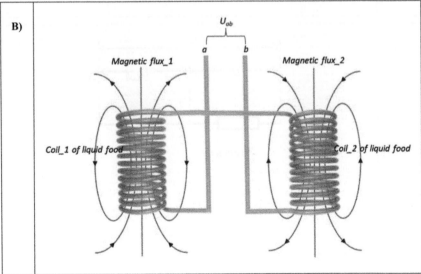

**FIGURE 10.3** Measurement system for the evaluation of the potential difference (or the voltage, $U_{ab}$) between two secondary coils or two secondary circuits. A) The coil system, and B) The circuit system.

In addition, two primary coils can electrically connect to one power source or different power sources. Differential potentials are generated between these coils or circuits because of changing in the level, frequency, and initial phase of primary voltages.

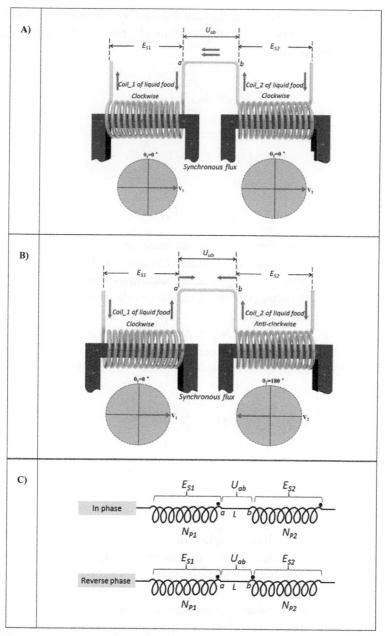

**FIGURE 10.4** A controllable $U_{ab}$ in liquid sample under the fluxes with different magnetic densities. (A) In-phase configuration, (B) Reverse-phase configuration, and (C) Equivalent circuit of these configurations.

When the voltages ($E_{S1}$ and $E_{S2}$) are induced by various fluxes in liquid sample simultaneously, transient potentials ($V_a$ and $V_b$) and controllable voltage ($U_{ab}$) are as follows:

$$V_a \propto U_{S1} \approx E_{S1} = U_{P1} \frac{N_{S1}}{N_{P1}} \tag{3}$$

$$V_b \propto U_{S2} \approx E_{S2} = U_{P2} \frac{N_{S2}}{N_{P2}} \tag{4}$$

$$U_{ab} = \left| V_a - V_b \right| \tag{5}$$

where, $V_a$, and $V_b$ are the transient potentials at terminal $a$ and $b$ of the secondary coils $1$ and $2$ or the secondary circuits $1$ and $2$, respectively. $U_{S1}$ and $U_{S2}$ are the terminal voltages at the secondary coils $1$ and $2$, $E_{S1}$ and $E_{S2}$ are the induced voltages at the secondary coils $1$ and $2$, $N_{P1}$ and $N_{P2}$ are the turns of the primary coils $1$ and $2$, respectively.

Controlling $E_{S1} = E_{S2}$ (or $N_{P1} = N_{P2}$) or $E_{S1} \neq E_{S2}$ (or $N_{P1} \neq N_{P2}$), the $U_{ab}$ is obtained in these systems. It can be used for evaluating the physico-chemical properties of liquid sample. In theory, the IEF ($E_{ab}$) is calculated as follows:

$$E_{ab} = \frac{U_{ab}}{L} = \frac{\left| V_a - V_b \right|}{L} \tag{6}$$

The measured $U_{ab}$ is shown in Figure 10.5. It is evident that the voltage between two equipotential cells of liquid food is improved with an increase in the excitation voltage, temperature, and frequency. The in-phase $U_{ab}$ is higher than the reverse-phase $U_{ab}$.

In practice, the spiral sample coil, namely the secondary coil, is arranged in a glass chamber. A constant-temperature circulating bath is connected to the chamber to maintain the temperature of the measurement. The insulating material such as mineral oils, vegetable oils, and synthetic esters, is used as the constant-temperature fluid, which can prevent the flux loss. The established system is shown in Figure 10.6. And, the evaluation of output voltages and IEFs is operable.

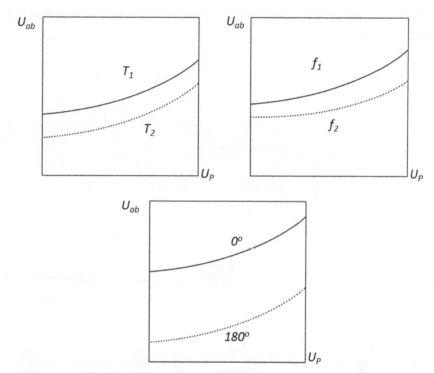

**FIGURE 10.5**   $U_{ab}$ influenced by different conditions. A) Influence of temperature as the change of excitation voltage; B) Influence of frequency as the change of excitation voltage; C) Influence of phase difference between two induced voltages as the change of excitation voltage (0°, in-phase; 180°, reverse-phase).

### 10.2.3   SYSTEM DESIGN PRINCIPLE

In a single-phase system, the secondary circuit of liquid food is considered a load, which is also theoretically a coil conductor. When the two terminals of the secondary coil of the sample are directly connected, in practice, the food is an equipotential cell. Therefore, the production of IEFs in foods is a key issue in the design of the system. The continuous flow system provides a necessary guarantee for the nonthermal treatment of food fluid. Various coil configurations under differential magnetic fluxes can also be exploited for generating IEFs (Figure 10.7). Although a weak electric current run through liquid sample, "coil loss" and "load loss" cannot be ignored under short-circuit, especially during long-term treatment. These losses cause potentially "hot spots" in electrolyte system, which depends on the ion current density.

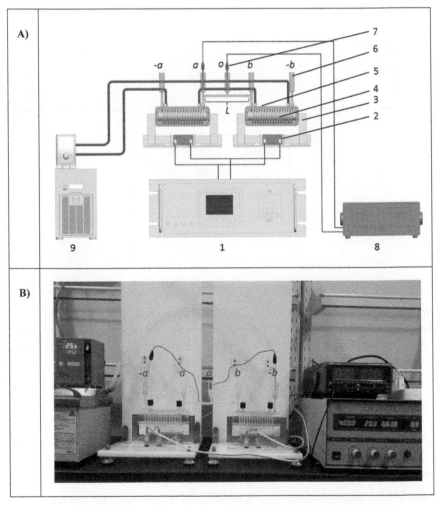

**FIGURE 10.6**   IEF measurement system. (A) Instrumental chain ["1) AC power source, 2) primary coil, 3) magnetic circuit, 4) spiral sample coil (interlayer glass spiral tube), 5) constant-temperature chamber, 6) glass link hole (measuring points), 7) platinum plate electrodes, 8) voltmeter, and 9) constant-temperature circulating bath"], and (B) Photograph of the system.

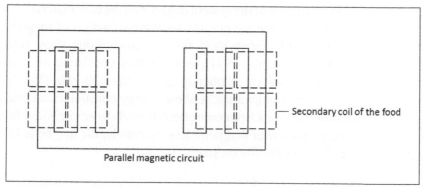

**FIGURE 10.7**   Eight sample coils arranged in a parallel magnetic circuit.

Food or other electrolyte system is used as the conductor and the load simultaneously, so these experimental transformer systems are actually working abnormally (Yang et al., 2016, 2017). An increase in the temperature of the core is inevitable during the treatment, and thus a chamber containing cooling water is required outside of the core (Figure 10.8).

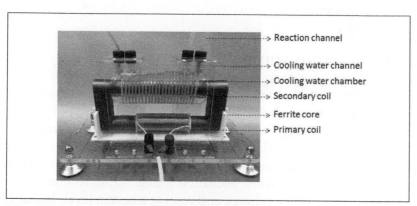

**FIGURE 10.8**   IEF reactor for circulating cooling water.

Please note that the structure of the system is only referred to as the transformer. The ionic current in the secondary circuit of liquid sample does not induce an EMF in the primary coil, and thus, there is no increase in the primary current during the operation. The phenomenon is caused by the migration mode of free ions, in which no magnetic effect is observed in electrolyte systems under the voltage.

The operating frequency of the system depends on the core material. The cores are assembled as magnetic circuits for the conduction of magnetic flux. The cores made of silicon steel sheet or amorphous alloy could operate at 50–1000 Hz (Narita, 1991), whereas they consisting of ferrite block work at an intermediate frequency greater than 20 kHz (Winder, 2017). However, increasing awareness about energy consumption causes the use of core steel materials from non-oriented steels to amorphous alloy or oriented steels with a lower magnetic loss. Normally, the magnetic loss is greater for ferrite than for silicon steel.

## 10.3 PROCESSING FACTORS

The IEF treatment is commonly performed in a continuous flow system owing to the multi-fluctuations pipeline configuration (Yang et al., 2017). Food fluid is subjected to the IEFs induced by various fluxes within an arrangement of secondary coils (or the reactors). The potential nonthermal effect on food fluids depends on excitation voltage, frequency, coil configuration, winding direction of the coil, as well as the conductivity and fluid impedance.

### 10.3.1 EXCITATION VOLTAGE

The excitation voltage, namely the primary voltage, is controlled by a power supply. An increased excitation voltage means that high IEF in liquid sample is induced. But, the thermal effect of the core increases with the increase of magnetic flux density. Although a higher cross-sectional area of the core can load high magnetic flux, it results in higher costs and bulkiness of the system. When excitation voltage, cross-sectional area of the core, and turns of the primary coil reset at a certain level, higher IEF can be induced by using fewer secondary coil turns.

### 10.3.2 FREQUENCY

The operating frequency depends on the material of the core. In practice, the core works in an abnormal condition during IEF treatment because food liquid (or electrolyte system) acts as the secondary circuit without a magnetic effect. A magnetic circuit with low core loss is preferred (Mack, 2005), and the system operating at high frequency can obtain a high potential difference between secondary coils or secondary circuits of liquid food.

### 10.3.3  COIL CONFIGURATION

Multi-series coils arranged in different positions of the magnetic circuit cause differential IEFs. The instantaneous polarity of induced voltage depends on the winding direction of the coil. Two secondary coils connected in series with the same directions can increase IEF strength under synchronous flux. The system has a compact structure when sample coils are located on multiple parallel magnetic circuits (Figure 10.9).

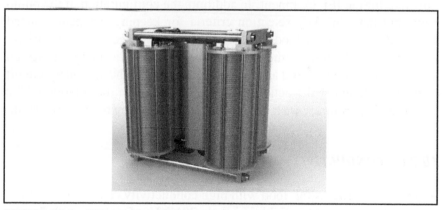

**FIGURE 10.9**   IEF system with four sample coils arranged in a 4-parallel magnetic circuit.

### 10.3.4  TEMPERATURE

The interaction effect between alternating electric field and temperature on pectin-methylesterase was investigated by a special-made device (Samaranayake and Sastry, 2016b). A significant nonthermal effect of the field on microorganisms and enzymes was observed, and the effect depended on the field strength and temperature (Meneses et al., 2013). Thus, exploring the applications at a controlled temperature is important for observing the nonthermal effect during electric field treatment. For keeping the temperature of the secondary coils or secondary circuits of the liquid food, a cooling fluid chamber is essential. To avoid the flux loss in conductive cooling fluid, insulating fluid is used as the cooling fluid to maintain the processing temperature.

### 10.3.5 FLOW IMPEDANCE

A high viscosity of the food fluid and a long pipeline result in high pump pressure needed for the IEF treatment. The processing time depends on the flow rate and the inner diameter of sample coil. Various food fluids acting as secondary coil and secondary circuit have differential flow impedance. High flow rate resulting in low fluid impedance under the electric field could improve the efficiency of the treatment. But there is a threshold of flow rate for the IEF treatment. In addition, the configuration of the pump body and parts are key selection criteria for keeping magneto-induced electrical effect on the food fluid. Otherwise, the effect of the electricity will terminate in flow path. For instance, a peristaltic pump causes an unsteady flow of the liquid and intermittently breaks the secondary circuit of the sample under the action of these fluxes due to the rotation of the pump head; therefore, a piston pump is preferred during the IEF treatment.

### 10.3.6 CONDUCTIVITY

As in electro-processing, food with high conductivity can be treated more effectively under electric fields (Li et al., 2017; Van Der Sman, 2017). The products are subjected to IEFs in tubes between secondary coils under different fluxes. The current density is influenced by the field strength, the inner diameter of the tube, and the conductivity of the sample. Thus, the treatment of liquid food with high conductivity results in a high-density energy input and short treatment times.

### 10.4 APPLICATIONS IN FOOD PROCESSING INDUSTRIES

IEF for nonthermal processing of food prevents adverse electrochemical reaction and electrode corrosion because of the avoidance of charged electrodes. But, the treatment of biomass granules can easily block the pipelines, and can be deposited in the spiral sample coils in the system (Yang et al., 2017). Thus, IEF processing is suitable for liquid foods containing juice, egg wash, vinegar, milk, and sauce, *etc.* The potential applications of this technique include modification of protein, sterilization, and inactivation of enzymes. In our previous works, an AC measurement system was established for the assessment of physicochemical properties of the liquid samples containing foods or carbohydrate polymers *via* the adjustment of excitation voltage, frequency, and potential points (Jin et al., 2015; Li et al., 2017).

During the processing, the food fluid of the secondary circuit can't connect to the ground for preventing significant thermal effect because of short-circuits current. But, connecting the secondary coil of the fluid through electrolytic bridge (the sample with different impedance) is also a feasible option for obtaining high potential difference. According to litera-tures, IEF-assisted hydrolysis of the biomass during long-term treatment may be caused by the hot spots (Li et al., 2016, 2017; Wu et al., 2017; Zhou et al., 2017). The pipelines between secondary coils or secondary circuits are the processing area; multi-parallel coils connected to a narrow pipeline can increase the induced current density in the sample. In addition, various waveform of the IEF can be adjusting by differential potentials under asynchronous flux.

## 10.5  SUMMARY AND CONCLUSION

At present, there is a growing demand for nonthermal food technology. IEF is developed based on magnetoelectric coupling with liquid food or electrolyte system acting as the secondary coil or circuit. According to *Faraday's* effect and *Ampere's law,* IEF emerges in liquid sample as a result of different magnetic fluxes in magnetic circuits, resulting in thermal under closed-loop or nonthermal effect under the open-loop. Differential fluxes are used as a stimulus rather than charged electrode avoiding undesirable electrochemical reactions, electrode corrosion, and metal contamination during this electrical treatment. The key to the design of the IEF system is the structural and electromagnetic calculation of the magnetic circuit, the design of the flow pipeline, and the overall arrangement of the potential difference area.

According to *Faraday's law* of Induction, the IEF in liquid food or electrolyte system is developed based on the magnetoelectric coupling. Induced voltages are applied and emerge in secondary coils of liquid sample by controlling variable fluxes. A secondary circuit of the food is an equipotential cell, based on the working principle of isolation transformer under short-circuit. When magnetic fluxes in two coils of the sample are different, the IEF would be produced, resulting in thermal or nonthermal effect, eventually affecting the processing. Compared to other electrotechnologies, differential fluxes are used as a stimulus rather than charged electrode, avoiding undesirable electrochemical reactions, electrode corrosion, and metal contamination during the electric field process. This chapter introduces the IEF principle, technical aspects, processing parameters, and potential applications. The contents highlight that IEF is an alternative electrotechnology for nonthermal processing.

## KEYWORDS

- **asynchronous flux**
- **coil configuration**
- **coil loss**
- **conductivity**
- **electroanalysis**
- **electrodes**
- **electrolyte system**
- **electromagnetic induction**
- **equipotential cell**
- **excitation voltage**
- **fluid impedance**
- **food fluid**
- **frequency**
- **impedance**
- **induced electric field**
- **induced voltage**

- **in-phase configuration**
- **joule heating**
- **liquid sample**
- **load loss**
- **magnetic circuit**
- **magnetic flux**
- **magnetic loss**
- **magnetic path length**
- **magnetomotive force**
- **nonthermal process**
- **ohmic law**
- **open-circuit**
- **phase angle**
- **phase difference**
- **potential difference**
- **primary current**
- **primary voltage**
- **reverse-phase configuration**
- **short-circuit**
- **synchronous flux**
- **terminal voltage**
- **transformer**
- **transient potential**
- **winding direction**

## REFERENCES

Bahill, A. T., & Baldwin, D. G., (2007). Describing baseball pitch movement with right-hand rules. *Computers in Biology and Medicine*, *37*(7), 1001–1008.

Caminiti, I. M., Noci, F., Morgan, D. J., Cronin, D. A., & Lyng, J. G., (2012). The effect of pulsed electric fields, ultraviolet light or high intensity light pulses in combination with manothermosonication on selected physico-chemical and sensory attributes of an orange and carrot juice blend. *Food and Bioproducts Processing*, *90*(3), 442–448.

Chandrasekaran, S., Ramanathan, S., & Basak, T., (2013). Microwave food processing—A review. *Food Research International*, *52*(1), 243–261.

Dalvi-Isfahan, M., Hamdami, N., Le-Bail, A., & Xanthakis, E., (2016). The principles of high voltage electric field and its application in food processing: A review. *Food Research International, 89*, 48–62.

Dermesonlouoglou, E., Zachariou, I., Andreou, V., & Taoukis, P., (2016). Effect of pulsed electric fields on mass transfer and quality of osmotically dehydrated kiwifruit. *Food and Bioproducts Processing, 100*, 535–544.

El Darra, N., Rajha, H. N., Ducasse, M. A., Turk, M. F., Grimi, N., Maroun, R. G., Louka, N., & Vorobiev, E., (2016). Effect of pulsed electric field treatment during cold maceration and alcoholic fermentation on major red wine qualitative and quantitative parameters. *Food Chemistry, 213*, 352–360.

Elez-Martínez, P., Sobrino-López, Á., Soliva-Fortuny, R., & Martín-Belloso, O., (2012). Chapter 4: Pulsed electric field processing of fluid foods A2. Cullen, P. J. In: Tiwari, B. K., & Valdramidis, V. P., (eds.), *Novel Thermal and Nonthermal Technologies for Fluid Foods* (pp. 63–108). San Diego: Academic Press, USA.

Fryer, P., (1995). Electrical resistance heating of foods. *In New Methods of Food Preservation* (pp. 205–235). Springer, USA.

Goullieux, A., & Pain, J. P., (2005). Ohmic heating 18. *Emerging Technologies for Food Processing* (p. 469).

Heathcote, M. J., (2007). 1 – Transformer theory. In: *J and P Transformer Book* (13th edn., pp. 1–13). Oxford: Newnes, UK.

Jaeger, H., Roth, A., Toepfl, S., Holzhauser, T., Engel, K. H., Knorr, D., Vogel, R. F., Bandick, N., Kulling, S., & Heinz, V., (2016). Opinion on the use of ohmic heating for the treatment of foods. *Trends in Food Science and Technology, 55*, 84–97.

Jin, Y., Yang, N., Duan, X., Wu, F., Tong, Q., & Xu, X., (2015). Determining total solids and fat content of liquid whole egg products via measurement of electrical parameters based on the transformer properties. *Biosystems Engineering, 129*, 70–77.

Khabdullin, A., Khabdullina, Z., Khabdullina, G., & Tsyruk, S., (2017). Development of a software package for optimizing the power supply system in order to minimize power and load losses. *Energy Procedia, 128*, 248–254.

Kuson, P., & Terdwongworakul, A., (2013). Minimally-destructive evaluation of durian maturity based on electrical impedance measurement. *Journal of Food Engineering, 116*(1), 50–56.

Li, D., Guo, L., Yang, N., Zhang, Y., Jin, Z., & Xu, X., (2017). Evaluation of the degree of chitosan deacetylation via induced-electrical properties. *RSC Advances, 7*(42), 26211–26219.

Li, D., Yang, N., Jin, Y., Guo, L., Zhou, Y., Xie, Z., Jin, Z., & Xu, X., (2017). Continuous-flow electro-assisted acid hydrolysis of granular potato starch via inductive methodology. *Food Chemistry, 229*, 57–65.

Li, D., Yang, N., Jin, Y., Zhou, Y., Xie, Z., Jin, Z., & Xu, X., (2016). Changes in crystal structure and physicochemical properties of potato starch treated by induced electric field. *Carbohydrate Polymers, 153*, 535–541.

Li, D., Yang, N., Zhou, X., Jin, Y., Guo, L., Xie, Z., Jin, Z., & Xu, X., (2017). Characterization of acid hydrolysis of granular potato starch under induced electric field. *Food Hydrocolloids, 71*, 198–206.

Mack, Jr. R. A., (2005). Chapter 9 – transformer selection. In: *Demystifying Switching Power Supplies* (pp. 263–275). Burlington: Newnes, CA.

Mahato, D. K., Verma, D. K., Billoria, S., Kopari, M., Prabhakar, P. K., Ajesh, K. V., Behera, S. M., & Srivastav, P. P., (2017). Applications of nuclear magnetic resonance in food processing and packaging management. In: Meghwal, M., & Goyal, M. R., (eds.), *Developing Technologies in Food Science Status, Applications, and Challenges* (Vol. 7). as part of book series on "Innovations in Agricultural and Biological Engineering" Apple Academic Press, USA.

Marra, F., Zhang, L., & Lyng, J. G., (2009). Radio frequency treatment of foods: Review of recent advances. *Journal of Food Engineering, 91*(4), 497–508.

Martín-Belloso, O., Soliva-Fortuny, R., Elez-Martínez, P., Robert, M. F., A., & Vega-Mercado, H., (2014). Chapter 18: Nonthermal processing technologies A2. Motarjemi, Yasmine. In: Lelieveld, H., (ed.), *Food Safety Management* (pp. 443–465). San Diego: Academic Press, USA.

Meneses, N., Saldaña, G., Jaeger, H., Raso, J., Álvarez, I., Cebrián, G., & Knorr, D., (2013). Modeling of polyphenoloxidase inactivation by pulsed electric fields considering coupled effects of temperature and electric field. *Innovative Food Science and Emerging Technologies, 20*, 126–132.

Nakonieczna, A., Paszkowski, B., Wilczek, A., Szypłowska, A., & Skierucha, W., (2016). Electrical impedance measurements for detecting artificial chemical additives in liquid food products. *Food Control, 66*, 116–129.

Narita, K., (1991). Silicon steel sheets. In: *Magnetic Materials in Japan* (pp. 107–212). Oxford: Elsevier.

Ohshima, T., Tanino, T., Kameda, T., & Harashima, H., (2016). Engineering of operation condition in milk pasteurization with PEF treatment. *Food Control, 68*, 297–302.

Pataro, G., Barca, G. M., Pereira, R. N., Vicente, A. A., Teixeira, J. A., & Ferrari, G., (2014). Quantification of metal release from stainless steel electrodes during conventional and pulsed ohmic heating. *Innovative Food Science and Emerging Technologies, 21*, 66–73.

Regier, M., Knoerzer, K., & Schubert, H., (2017). Chapter 3: Determination of the dielectric properties of foods. In: *The Microwave Processing of Foods* (2nd edn., pp. 44–64). Woodhead Publishing.

Samaranayake, C. P., & Sastry, S. K., (2016a). Effect of moderate electric fields on inactivation kinetics of pectin methylesterase in tomatoes: The roles of electric field strength and temperature. *Journal of Food Engineering, 186*, 17–26.

Samaranayake, C. P., & Sastry, S. K., (2016b). Effects of controlled-frequency moderate electric fields on pectin methylesterase and polygalacturonase activities in tomato homogenate. *Food Chemistry, 199*, 265–272.

Saslow, W. M., (2002). Chapter 12 – Faraday's law of electromagnetic induction. In: *Electricity, Magnetism, and Light* (pp. 505–558). San Diego: Academic Press.

Stancl, J., & Zitny, R., (2010). Milk fouling at direct ohmic heating. *Journal of Food Engineering, 99*(4), 437–444.

Suresh, Y., & Panda, A. K., (2016). Investigation on stacked cascade multilevel inverter by employing single-phase transformers. *Engineering Science and Technology, an International Journal, 19*(2), 894–903.

Timmermans, R., Groot, M. N., Nederhoff, A., Van Boekel, M., Matser, A., & Mastwijk, H., (2014). Pulsed electric field processing of different fruit juices: Impact of pH and temperature on inactivation of spoilage and pathogenic micro-organisms. *International Journal of Food Microbiology, 173*, 105–111.

Tomczuk, B., & Koteras, D., (2011). Magnetic flux distribution in the amorphous modular transformers. *Journal of Magnetism and Magnetic Materials, 323*(12), 1611–1615.

Van Der Sman, R., (2017). Model for electrical conductivity of muscle meat during Ohmic heating. *Journal of Food Engineering, 208,* 37–47.

Vega-Mercado, H., Martin-Belloso, O., Qin, B. L., Chang, F. J., Góngora-Nieto, M. M., Barbosa-Cánovas, G. V., & Swanson, B. G., (1997). Nonthermal food preservation: Pulsed electric fields. *Trends in Food Science and Technology, 8*(5), 151–157.

Villén, M. T., Letosa, J., Nogués, A., & Murillo, R., (2013). Procedure to accelerate calculations of additional losses in transformer foil windings. *Electric Power Systems Research, 95,* 85–89.

Watanabe, T., Ando, Y., Orikasa, T., Shiina, T., & Kohyama, K., (2017). Effect of short time heating on the mechanical fracture and electrical impedance properties of spinach (*Spinacia oleracea* L.). *Journal of Food Engineering, 194,* 9–14.

Winder, S., (2017). Chapter 12 – Magnetic materials for inductors and transformers. In: *Power Supplies for LED Driving* (2nd edn., pp. 241–247). Newnes, USA.

Wu, F., Jin, Y., Li, D., Zhou, Y., Guo, L., Zhang, M., Xu, X., & Yang, N., (2017). Electrofluid hydrolysis enhances the production of fermentable sugars from corncob via in/reverse-phase induced voltage. *Bioresource Technology, 234,* 158–166.

Yang, N., Jin, Y., Li, D., Jin, Z., & Xu, X., (2017). A reconfigurable fluidic reactor for intensification of hydrolysis at mild conditions. *Chemical Engineering Journal, 313,* 599–609.

Yang, N., Jin, Y., Wang, H., Duan, X., Xu, B., Jin, Z., & Xu, X., (2015). Evaluation of conductivity and moisture content of eggs during storage by using transformer method. *Journal of Food Engineering, 155,* 45–52.

Yang, N., Zhang, N., Jin, Y., Jin, Z., & Xu, X., (2017). Development of a fluidic system for efficient extraction of mulberry leaves polysaccharide using induced electric fields. *Separation and Purification Technology, 172,* 318–325.

Yang, N., Zhu, L., Jin, Y., Jin, Z., & Xu, X., (2016). Effect of electric field on calcium content of fresh-cut apples by inductive methodology. *Journal of Food Engineering, 182,* 81–86.

Yu, X., Xi, W., Liu, Z., Kuang, Y., Li, H., Fu, X., Liu, X., Xu, W., Song, Y., & Wu, S., (2017). Design of a 150T pulsed magnetic field generator device. *Fusion Engineering and Design, 121,* 265–271.

Zhou, Y., Jin, Y., Yang, N., Xie, Z., & Xu, X., (2017). Electrofluid enhanced hydrolysis of maize starch and its impacts on physical properties. *RSC Advances, 7*(31), 19145–19152.

# Index

## α

α-phosphatase, 5
α-glucosidase, 219

## β

β-carotene, 47, 71, 73, 87, 231, 234
β-galactosidase, 218
β-glucosidase, 236
β-glucuronidase, 247
β-lactoglobulin, 4, 241
β-phosphatase, 5

## γ

γ-glutamyl group, 9
γ-glutamyltransferase, 2, 3, 20

## A

Abiotic factors, 78
Acerola, 97
Acid
   cysteine, 253
   phosphatise, 218
Activation energy, 45, 98, 99, 191
Actomyosin, 248
Aerobic
   bacteria, 127, 234
   conditions, 83
   degradation, 92
Air
   drying (AD), 43, 45, 47, 75, 97–99, 101,
      102, 162, 176, 178, 187–189
   microwave (AM), 97
   velocity, 45, 72, 74, 99, 160
Aldehydes, 84, 95, 236
Alkaline phosphatase (ALP), 2, 4–9,
   11–14, 19, 20, 241, 254
   thermal stability, 8
Alkaloids, 5
Alternative current (AC), 121, 259, 268, 273

Amine group, 83, 102
Amino acids, 9, 47, 83–85, 89–91, 243,
   244, 246
Annealing effect, 47
Annihilation, 208
Anode, 28
Antagonistic effect, 242
Anthocyanins, 39, 40, 42, 47, 74, 75, 77,
   87, 185, 221, 235
Antibiotics, 8
Antimicrobial
   activity, 3
   agents, 10
Antimutagenicity, 231
Antioxidant activity (AA), 38, 40–42, 47,
   52, 92, 220, 234
Antiviral properties, 10
Apple pomace (AP), 220
Aroma, 84, 85, 95, 96, 231, 235
Ascorbates, 234
Ascorbic acid, 8, 38, 39, 41, 42, 71, 82–84,
   91–94, 232, 234
   degradation, 82
Ascorbinase, 82
Asynchronous flux, 273, 274

## B

*Bacillus stearothermophilus*, 236
Bacterial
   growth, 52
   spores, 49, 220
Barosensitivity, 227
Benzaldehyde, 236
Benzothiazoilhydroxy-benzothiazole
   phosphate, 7
Bilharziasis, 117
Bioactive non-nutrient components, 85
Biotic factors, 78
Blanching, 35, 37–39, 54, 82, 88, 97, 178,
   227, 230, 232, 245

Brewer's yeast, 219
Brown pigment, 82
Browning, 39, 74, 81–84, 168, 233, 234

# C

Calcium infusion, 232
Calpains activity, 252, 254
Calpastatin, 252, 254
Caprine, 2, 11, 14, 16, 18, 19
    milk, 19
    origin, 14
Caramelization, 82
Carbohydrate, 1, 88, 89, 158, 226
    contents, 89
    polymers, 273
Carbon, 152, 220, 245
    dioxide (CO2), 220, 245
Cardiovascular diseases, 90
Carotenoids, 42, 51, 84, 87, 161, 231
Casein
    micelles, 3
    molecules, 238
Catalase, 3
Catalytic activity, 4
Catecholase, 82
Cathode, 28
Cavitation, 163, 180, 199–222
    generation mechanism, 200
        hydro thermodynamic (HTD) cavita-
            tion, 203
        hydrodynamic cavitation (HD), 98,
            101, 200–203, 205, 208, 211,
            215–221
    inception number, 202, 209
    intensity, 201, 202, 205, 207, 209, 210,
        215, 216, 218, 221, 222
    number, 201, 202, 207–214, 218, 219, 222
    system, 215
    techniques, 221
    threshold, 215
    yield, 213, 214
Cavitational
    activity, 215
    yield, 213
Cavity
    collapse, 207

generation, 202, 207, 216
    initiation, 214
Cell
    disruption, 212, 216, 221, 222
    membrane, 50, 120, 121, 182
        rupture, 50
    rupture, 50
    walls, 177, 189, 219
Cellular
    glutathione, 9
    inactivation, 219
    materials, 50
    matrix, 41
    water, 41
Cellulose
    crystallization, 81
    membranes, 118
Cellulosic fibers, 90
Centrifugal force (CF), 178
Cheese proteolysis, 244
Chelators, 6
Chemical
    contaminants, 117, 127
    decomposition, 48
    drying, 73
Chemiluminescent, 6, 7
Chitosan, 118, 127
Chlorophyll, 84, 87, 231
Cholera, 117
Cholesterol, 86, 131, 132, 139, 140, 142,
    143, 149–152
    diffusivity coefficient, 149–152
Chronic biochemical disorders, 94
Coil
    configuration, 262, 267, 270, 271, 274
    loss, 267, 274
Coliform bacteria, 118, 121, 127
Collagen, 91, 246
Collapse pressure, 207, 209, 210, 213, 214,
    218, 219, 222
Colon
    cancer, 90
    parameters, 161, 167
Colony-forming unit (CFU), 121, 125,
    127, 240, 250
Colorimeter, 161
Colorimetric methods, 6, 7

Conductivity, 31, 35, 37, 40, 119, 127, 216, 270, 272, 274
Constant
   cell concentration, 219
   temperature fluid, 266
   velocity, 160
Convective
   drying (CD), 43–45, 73, 74, 95, 98, 157, 158, 160, 162, 164, 165, 168–170
   pre-drying (CPD), 47
Conventional
   drying systems, 43, 45, 54
   heating, 26, 27, 30, 37, 40, 54
Correlation coefficient, 166, 167
Cospores, 220
Covalent bonds, 226, 230
Critical
   cavity density, 218
   temperature, 132–134
Crude protein, 74
Crystallinity, 40
Cytoplasmic, 217
Cytosolic fraction, 247

**D**

Dairy products, 2, 8, 12, 175, 237, 255
Dehydration, 35, 53, 70, 97, 158, 163, 166, 167, 171, 176–179, 181, 182, 184, 185, 187–192, 227, 245, 259
Dehydroascorbic acid (DAA), 92
Denaturation, 237, 238, 241, 245
Deoxyribonucleic acid (DNA), 50, 117
Dephosphorylation, 5
Depressurization, 206, 235, 253
Destruction rate, 125, 127
Deterioration, 37, 41, 47, 51, 53, 78–80, 85
Dielectric
   characteristics, 26, 31
   constant, 26, 31–34, 55, 132
   loss, 26, 31–33, 55
      factor, 31–33, 55
   properties, 26, 31–33, 52, 55, 165, 170
   strength, 120
Dietary fibers, 90
Differential fluxes, 273
Diffusion coefficient, 73, 80, 135

Diffusional mechanism, 165
Diffusivity, 45, 77, 85, 98, 132, 135, 149–152, 166, 167, 170, 188, 191
Digital balance, 160, 161
Dioxide chlorine (ClO$_2$), 118
Diphenolase enzymes, 82
Dipole, 28, 31, 32, 55
   rotation, 28, 31, 55
Discoloration, 81
Disinfection, 117, 118, 125, 127, 200, 221
Divergence angle, 214
Diverticular diseases, 90
Domestic microwave, 160
Downstream section, 214
Drying
   constant, 73
   effects, 78
      carbohydrate content, 89
      chemical quality, 81
      color, 81
      fat content, 89
      fiber content, 90
      flavor, 84
      mineral content, 94
      nutritional quality, 88
      physical quality, 80
      phytochemicals, 85
      protein content, 90
      rehydration quality, 96
      sensory quality, 95
      shrinkage, 80
      texture and porosity, 80
      vitamin content, 91
   kinetics, 38, 46, 47, 158, 162, 163, 165, 170, 176
   overview, 71, 73, 77
   rate, 29, 34, 43, 45, 47, 49, 54, 72–74, 80, 81, 163, 165, 176, 178, 183

**E**

Effective
   diffusivity, 162, 166, 180, 183, 185, 187, 188
   moisture diffusivity, 46, 183, 191
Electric
   conductivities, 120

field, 26, 28, 33, 119–121, 123,
    259–262, 271, 272, 274
  intensity, 120
Electrical
  conductivity (EC), 2, 3, 7, 119, 120, 127
  field, 118–123, 125–127
Electricpotential difference, 260
Electroanalysis, 262, 274
Electrochemical reactions, 260, 272–274
Electrodes, 118, 119, 125, 127, 259, 260,
    262, 268, 273, 274
  corrosion, 260, 272–274
Electrolyte system, 268–270, 273, 274
Electrolytic bridge, 273
Electromagnetic
  calculation, 273
  coupling, 50
  energy, 26, 32
  field, 26, 27, 30, 31
    intensity, 26
  induction, 261, 274
  wavelength, 32
  waves, 25
Electromotive force (EMF), 260, 269
Electron-emitting cathode, 28
Electroporation, 50, 120–122
Electrotechnologies, 259, 274
Emission wavelength, 7
Empirical equations, 142, 152
Endogenous milk enzymes, 11
Energy
  dissipation, 26, 207, 209, 210
  intensive process, 42
*Enterobacteriaceae*, 250
*Enterococcus*, 250
Enzymatic
  activities, 14, 19, 82
  browning, 37, 82
  indicator, 5
  reactions, 47
  routes, 79
Enzyme inactivation, 9, 19, 203
Equilibrium, 84, 133, 162, 226, 230
  moisture content (EMC), 73
Equipotential cell, 262, 266, 267, 274
Equivalent circuits, 263
*Escherichia coli (E. coli)*, 51, 118, 121,
    122, 127, 212, 217, 218, 240, 251

Esters, 84, 266
Excitation voltage, 262, 266, 267, 270,
    273, 274
Experimental drying kinetics, 163
Extrinsic indicators, 5

**F**

Fat globules, 9, 14, 238, 244
Fatty acids, 85, 89, 133
Fecal coliform bacteria, 118, 127
Fermentation, 87, 240, 259
Field polarity, 28
Filamentous fungi, 232
First-order kinetic inactivation, 16
Flavanones, 234
Flavonoids, 8, 40
Flavonol, 40
Flow impedance, 272
Fluid
  impedance, 270, 272, 274
  velocity, 214
Fluorimetric methods, 6, 7
Fluorometric assays, 7
Folates, 234
Food
  Agricultural Organization (FAO), 10,
      89, 91
  dimension, 37
  Drug Administration (FDA), 227
  engineering, 170
  fluid, 267, 270, 272–274
  geometry, 52
  matrix, 4, 5
  pasteurization, 219
  preservation, 102, 175
  processing, 26, 27, 30, 70, 77, 78, 176,
      177, 200–222, 226, 227, 262
    application, 35
    HPP application, 230
    hydrodynamic cavitation (HD), 216
    hydro-thermodynamic cavitation, 220
    industries applications, 272
  quality, 71, 76, 102, 178, 192
  spoilage, 228, 255
  sterilization, 219
  thermal properties, 35

Free
  fatty acids, 133
  radicals, 216, 249
Freeze-drying (FD), 46–49, 55, 73–75, 77, 78, 94, 97, 100, 176
Frequency, 25–27, 31–33, 36, 37, 50, 53, 55, 159, 179, 180, 184, 206, 213, 260, 262, 264, 266, 267, 270, 271, 273, 274
Friction, 28
Frozen broccoli, 42
Fructose, 41, 84, 89, 177, 186
Functional foods, 87
Furosine, 4

**G**

Gelatinization, 35, 41, 47, 81
Gelation, 245
Genotype, 41
Geometric parameters, 207, 222
Glass transmission temperature, 47
Glassy matrix, 98
Globin, 237
Globule membrane, 3, 5, 9, 11
Glucose, 41, 84, 89, 177, 231
Glucosinolates, 40
Glutamic acid, 84
Glutamyl
  cycle, 9
  group, 9
  transferase, 242
Glutamyltransferase (GT), 2–5, 9, 14–16, 19, 20
  enzymatic activity, 14
Glutathione, 9
Glycoprotein, 9, 10
Glycylglycine, 9
Gram
  negative
    bacteria, 237
    cells, 219
  positive bacteria, 237
Greek yogurt, 220
Guava slices, 94, 180

**H**

Hammer effect, 216

Heat
  transfer, 26, 29, 37, 39, 42, 48, 53, 72, 165, 170, 180, 183, 191
  mechanism, 29
  treatments, 1, 2, 4, 9, 17
  effect, 4
Helical coils, 51
Hemicellulose, 90
High
  hydrostatic
    pressure, 178
    processing (HHP), 226
  intensity pulsed electric field, 259
  pressure
    homogenizer (HPH), 203, 217
    processing (HPP), 200, 225–242, 244–248, 250, 251, 253–255, 257
    thawing, 53
  temperature short time (HTST), 1, 6, 9, 15
Homogeneous drying, 160
Homogenization, 217, 220
Homogenizer valve design, 204
Host-defense systems, 10
Hot
  air (HA), 38, 43–45, 47, 53, 70, 72, 93, 96, 97, 100, 184
    convective drying (HACD), 44, 47
    drying (HAD), 45, 48, 70, 95, 96
  water (HW), 37, 38, 42, 51, 52, 176
    blanching (HWB), 38, 39, 176
Humidity hot air impingement blanching (HHAIB), 38
Hybrid drying (HD), 98, 101, 102
Hydration, 47, 96
Hydro thermodynamic cavitation, 222
Hydrodynamic cavitation (HD), 199, 200, 202, 205–207, 209, 215, 223
Hydrogen peroxide, 10, 51
Hydrolysis, 7, 41, 92, 273
Hydro-peroxides, 89
Hydrophobic interactions, 230
Hydro-thermodynamic cavitation, 220
Hydroxymethylfurfural, 4
Hydroxyquinoline, 7
Hyperbaric pressure, 226
Hypertonic solution (HS), 177, 180
Hypothiocyanite, 10

## I

Immunochemical methods, 6
Impedance, 270, 272–274
Inactivation
  kinetics, 14, 19, 242
  mechanisms, 11
Indicator enzyme, 4, 5
Induced electric field (IEF), 259, 260, 262, 266, 268–274
  processing theory, 261
    inductive methodology, 261
    potential difference, 262
    system design principle, 267
Induced voltage, 260, 261, 263, 266, 267, 271, 274
Infrared blanching (IRB), 38
Initial freezing point, 52
Inlet
  fluid temperature, 220
  pressure, 201, 207, 208
Inositol hexakisphosphate (Insp6), 86
In-phase configuration, 263, 275
*In-situ* evaluation, 4
International Dairy Federation (IDF), 7, 10, 20
Intrinsic indicators, 5
Invertase, 217, 219
*In-vitro* starch digestibility (IVSD), 220
Ionic
  conduction, 28
  interactions, 230
  polarization, 28, 55
Irradiation, 51, 175, 227, 245
Isoelectric point, 9
Isomers, 82
Isostatic
  principle, 226, 230
  rule, 226
Isozymes, 5

## K

Kaempferol, 40
Ketones, 84, 89
Kinetic comparative studies, 11
  alkaline phosphatase (ALP) thermal
    inactivation, 11

ALP activity, 11
ALP thermal inactivation mechanism, 12
GT thermal inactivation, 14
  GT activity, 14
  GT thermal inactivation mechanism, 15
LP thermal inactivation, 17
  LP activity, 17
  LP thermal inactivation mechanism, 18

## L

Lactation
  cycle, 5, 14
  period, 9
Lactic acid bacteria, 220, 250
*Lactobacillus*, 132, 152, 247
Lactoperoxidase (LP), 2–6, 10, 11, 17–20
Lactose, 8, 14
Lactulose, 4
Lipid, 14, 85, 89, 120, 158, 250
  globules membranes, 14
  peroxidation, 249
Lipoxygenase, 37
Liquid
  density, 202
  flow velocity, 202
  foods, 200, 219, 228, 261, 273
  physicochemical properties, 215
  sample, 260–262, 265–267, 269, 270, 273–275
  velocity, 201, 202
  water volume, 80
Load loss, 260, 267, 275
Local turbulence, 203
Loss tangent, 31, 33
Low
  electric field, 127
  pressure hydrodynamic cavitation (HD) reactor, 205
Lumry-Eyring model, 13
Lutein, 87
Lycopene, 87, 157, 158, 169, 170, 231, 236
*Lycopersicon esculentum*, 158
Lysosomal membranes, 247
Lysosomes, 247

Lysozyme, 3
L-γ-glutamyl-p-nitroanilide, 9

# M

Magnetic
  circuit, 260, 261, 268–271, 273, 275
  field, 261
    coupling, 50
  flux, 260–262, 267, 270, 273–275
  loss, 270, 275
  path length, 275
Magnetoelectric coupling, 273, 274
Magnetomotive force (MMF), 261, 275
Magnetron, 28, 29, 55, 161
  power, 29
Maillard
  browning, 83, 89
  reaction, 82, 83, 85
Maipo-ecotype artichokes, 38
Mass transfer, 26, 29, 30, 48, 70, 71, 143,
    152, 163, 170, 175, 177, 180, 191, 192,
    230
  coefficient, 143, 152, 191
  mechanism, 29
Mathematical modeling, 122, 127, 149, 171
Meat, 48, 132, 140, 149, 152, 175, 237,
    245–248, 250–252, 254, 255
  products, 237, 245, 250, 255
  tenderization, 237, 247, 251, 254
Mechanical drying, 75
Melanin, 82
Mesophiles, 233
Mesophilic flora, 250
Metabolic processes, 94
Methylation, 232
Micro/trace minerals, 94
Microbes, 227, 241, 245
Microbial
  cells, 50, 226, 228
    disruption, 216
  destruction, 233, 240, 246
  growth, 51, 53, 54, 69, 95, 246, 247
  inactivation, 48, 50, 219, 237, 251
  lethality, 237
  load, 38, 53, 226, 228, 240, 250
  reduction, 237, 241

  safety, 1, 237
  thermal-resistance, 237
Microbiological
  characteristics, 236
  criteria, 4, 227
  growth, 71
  platforms, 2
  quality, 10, 233, 246, 250
Microflora, 240
Microorganisms, 1, 4, 9, 49, 50, 52, 70, 76,
    117, 118, 120–123, 125, 127, 206, 219,
    220, 228, 240, 244, 271
Microstructure, 177, 183, 184, 220
Microwave (MW), 25–35, 37, 39–43,
    45–55, 70, 75, 85, 96–98, 100, 101,
    157–161, 164–166, 168, 170, 171, 176,
    178, 189
  assisted
    air drying (MWAD), 43, 44
    drying (MWD), 45, 55, 75
    extraction, 54, 55
    freeze-drying (MWFD), 48, 49, 55
    pasteurization, 55
    spouted bed (MWSB), 39
    sterilization, 55
    tempering/thawing, 55
    vacuum drying (MWVD), 44–47, 55
  blanching (MWB), 37–39, 55
  chamber, 160, 161
  circulated water combination (MCWC),
    52
  convective drying, 159
  cooking, 39, 40, 41
  density, 160
  drying (MWD), 34, 43–47, 75, 87, 94,
    97, 158, 160, 164, 165, 171
  energy, 30, 34, 43, 51, 53, 160, 164
  heating, 26, 27, 30, 39, 43, 50–52, 54
    affecting factors, 30
    baking and cooking, 39
    blanching, 37
    drying, 42
    pasteurization, 49
    sterilization, 52
    tempering/thawing, 52
  oven power measurement, 160
  overview, 27
    instrumentation, 28

working principle mechanism, 28
pasteurization, 51
pasteurizer, 51
power, 34, 39, 41, 43, 45
thawing, 53
vacuum-freeze dryer, 75
Microwaving, 40, 41, 42
Milk
  fat fraction, 11
  homogenization, 204
  plasma, 5, 14
Millard reaction, 102
Milling, 86, 87
Model decimal reduction time, 4
Moderate electric field (MEF), 120, 259
Moisture
  content (MC), 26, 30, 31, 33, 45, 48,
    52, 69, 72–75, 80, 81, 83, 158, 159,
    161–163, 165, 166, 175, 177, 183,
    192, 238, 239, 251
  retention, 238
Molecular
  friction, 26
  weight, 131, 133, 134, 150, 151, 226
Monocytogenes, 6, 237, 240, 249, 250,
  252, 253
*Moringa oleifera*, 73
Multihole orifice, 215
Mycobacterium tuberculosis, 6
Myofibril, 246
  fragmentation, 251
Myoglobin, 237
Myosin molecules, 245

### N

Nisin, 237, 240
Non-bovine
  milk, 12, 18–20
  origin, 2, 14, 17
  species, 12, 18
Non-polar compounds, 133
Nonthermal
  methods, 259
  process, 228, 261, 272, 274, 275
Non-uniform temperature, 30
Non-volatile components, 84

Novel
  methods, 132, 152
  thawing, 53
Nucleation, 216
Nucleotides, 5
Nutritional
  changes, 157
  property, 102

### O

Obesity, 90, 183
Ohmic
  heating (OH), 54, 215, 259, 260
  law, 275
  thawing, 53
Oil uptake (OU), 183
Open-circuit, 262, 275
Optimal drying condition, 170
Organic
  acids, 84, 244
  compounds, 133, 134
  molecule, 200
Organoleptic properties, 233
Orifice, 200–203, 205, 207–216, 221, 222
  diameter, 209
  geometrical parameters effect, 209
    flow area, 210
    orifice diameter, 209
    orifice shape, 211
    venturi geometrical parameters effect,
      214
  velocity, 210
Osmosis, 118, 127, 192
Osmotic
  dehydration (OD), 37, 70, 96, 97, 158,
    163, 165–167, 171, 176–182, 184,
    185, 187, 190–192, 227, 230
  pretreatment, 164
  process, 159, 163, 165, 169
  shock, 218
Ovine milk, 2, 11
Oxidation reaction, 82
  reduction reaction, 260
Oxidative degradation, 73
Oxygen radical absorbance capacity
  (ORAC), 221

# P

Page's model, 162, 164–166, 170
Parallel magnetic circuits, 271
Pascalization, 226
Pasteurization, 1, 5, 6, 8, 9, 12, 15, 17, 19,
  27, 35, 49–51, 54, 55, 200, 205, 220,
  221, 226, 227, 229, 230, 237, 241, 245,
  250
  method, 51, 237
Pathogenic
  bacteria, 6, 50
  microbes, 117
  microorganisms, 1
Pathogens, 49, 226, 227, 240, 251, 253
Pectin, 37, 81, 90, 181, 232, 235, 271
  degradation, 81
  enzymes, 37
  methylesterase (PME), 51, 234, 271
Penetration, 32–34, 39, 50, 71
Peptidoglycans layer, 219
Periodic pressure oscillation, 240
Periplasmic product, 212
Permittivity, 31, 32
Peroxidase (POD), 10, 20, 37, 39, 232,
  233, 235, 236
Persimmons, 86
Phase
  angle, 275
  difference, 263, 267, 275
Phenolase, 82
Phenolic, 39, 40, 185, 221
  acids, 40
  compounds (PCs), 47, 82, 84–86, 102
  content, 39
Phenoloxidase, 82
Phenols, 85
Phenyl phosphate, 7
Phenylalanine, 85
Pheophytin, 87
Phosphatase, 2, 3, 5, 212, 218, 247, 254
Phosphomonoesterases, 5
Phosphoric
  esters, 7
  monoesters, 7
Phosphorous, 86
*Phospohydrolases*, 3

Photodiode-array, 87
Physico-chemical
  characteristics, 232
  composition, 90
  properties, 47, 49, 101, 215, 235, 273
Phytate content, 86, 87
Phytochemicals, 41, 85
  composition, 42
Pigments, 38, 82, 84, 85, 87, 93, 226
Pipe diameter ratio, 212, 213
Plasmin, 241, 242, 244
Plasminogen, 242
Plate geometry, 205, 212
Platinum electrodes, 260
P-nitroaniline, 9
P-nitrophenyl phosphate, 5, 7
Polar molecules, 28
Polarity, 28, 29, 134, 271
Polarization, 28, 31
Poliomyclitis, 117
Polyethylene, 93
Polyphenol, 40, 82
  oxidase (PPO), 37, 39, 82, 86, 232, 233,
    235, 236
Polyphenolic compounds, 85, 221
Polytropic constant, 216
Porosity, 31, 47, 71, 81, 101
Potential
  applications, 273, 274
  difference, 50, 261, 262, 264, 271, 273, 275
  points, 273
Power density, 39, 47, 49
Pressure
  homogenization, 217, 218
  transmitting medium, 229
Pressurization, 229, 235–238, 242, 243,
  246–250, 252–254
Pretreatment, 100, 158, 159, 162–164,
  168–170, 176, 178, 183, 188
Primary
  current, 269, 275
  voltage, 264, 270, 275
Probiotic drinks, 245
Programmable logic controller (PLC), 160
Proliferation, 71, 117
Proportional integral derivative (PID), 160,
  228

Protein
 degradation, 238
 matrix, 244
Proteinase activity, 247, 249
Proteolysis, 241, 243, 244, 247, 253
Proteolytic activity, 253
*Pseudomonas*, 247
Pulse rate (PR), 159, 162, 164, 168, 169
Pulsed electric field (PEF), 176, 178, 200, 259, 260
Purple sweet potatoes (PSPs), 40, 41

## Q

Quantification, 6
Quercetin, 40
Quinolyl phosphate (QP), 7

## R

Radial bubble motion, 216
Radiofrequency, 176
Rancidity, 89, 248
Raw materials, 70
Reactivation capacity, 7
Rehydration, 45, 47, 48, 71, 74, 75, 77, 96–101, 177, 186, 192, 227, 245
 capacity, 97
 kinetics, 97, 101
 potential, 97
 rate, 101
 ratio (RR), 96, 98, 99, 186
Relative
 electrical permittivity, 32
 humidity (RH), 72
 permittivity, 26
Rennet coagulation time, 237, 242
Residual enzyme activity, 18
Resistant starch (RS), 40, 41
Response surface methodology (RSM), 191
Reverse
 osmosis (RO), 118, 125
 phase configuration, 263, 275
Reynold's number, 149
Rheological features, 220
Rhodamine B, 212, 213
 degradation, 212, 213
Ribonuclease, 3

Rigid cells, 81
Root mean square (RMS), 33
 error (RMSE), 123, 124

## S

*Saccharomyces cerevisiae*, 219
Sarcomere
 shrinkage, 246
 structure, 252
Sarcoplasmic proteins, 249
Secondary
 circuits, 262, 264, 266, 271–273
 lipid oxidation, 248
Semi-dried fruits, 171
Serine proteases, 253
Shelf-life (SL), 70–72, 77, 78, 102, 176, 192, 208, 213, 221, 226–228, 230, 231, 233, 234, 237, 240, 241, 252, 253, 255
Shock wave
 generation, 216
 reactors, 205, 206
Short-circuit, 262, 267, 273–275
Sialic acid, 8
Single-bubble cavitation, 199
Slit venturi, 208
Sodium chloride (NaCl), 8, 119, 159, 165, 169, 170, 177, 236, 248, 254
Solar
 dryer, 77
 drying, 77, 88, 93, 95
 energy, 78
Solid gain (SG), 169, 177, 180, 184, 186, 191
Soluble starch (SS), 41
Somatic cells, 2, 3, 14
Sonication, 97, 180, 181, 184, 187, 189, 190, 192, 216, 217
Sorghum flour (SF), 220
Spoilage bacteria, 10
Sponge effect, 163, 180, 181, 192
Standard error (SE), 166, 167
Staphylococci, 118
*Staphylococcus aureus*, 237
Starch granules, 47
Steam blanching (SB), 38, 39, 176
Steamer steaming, 41, 42

Sterilization, 1, 35, 49, 52, 54, 55, 118, 125, 127, 200, 222, 229, 235, 236, 259, 273
Steroid compounds, 131, 152
Stir-frying, 42
Sucrose, 39, 41, 84, 89, 177, 185, 186, 188–190, 246
Sulfhydryl groups, 8, 10
Sun drying, 76
Supercritical
    fluids, 132–135, 152
    CO2 (SC2-CO2), 132–135, 137–143, 149–152
        benefits, 134
        cholesterol diffusivity, 151
        cholesterol removal, 139
        cholesterol solubility, 142
        compressibility coefficient, 138
        drawbacks, 134
        physical properties, 134
        state, 132
    region, 132
Superoxide dismutase, 2, 3, 10
Surface
    browning, 54
    tension, 134, 215
Survival microorganism ratio, 122
    mathematical modeling, 122
Suspension temperature, 204
Synchronous flux, 263, 271, 275

**T**

Tactile handling, 239
Temperature
    gradient, 34
    indicators, 4
Terminal voltage, 266, 275
Texturization, 245
Thermal
    conductivity, 35
    diffusion, 27
    diffusivity, 35
    inactivation, 2, 4, 13, 15, 18, 242
    resistance characteristics, 4
    sensitivity values, 15
    softening rate, 232
    stability, 6, 18
    treatments, 226

Thermalization, 1
Thermodynamic principle, 226
Thermo-physical action, 71
Thiosulfinate, 74
Tilapia fillets, 254
Time temperature indicators (TI), 2, 4, 5
    alkaline phosphatase (ALP), 5
        ALP activity determination, 6
        importance, 6
        limitations, 7
    γ-glutamyltransferase (GT), 9, 10
        importance, 9
        lactoperoxidase (LP) in dairy industry, 10
        LP system reaction mechanism, 10
Titratable acidity values, 220
Total
    dietary fiber (TDF), 90, 220
    phenolic
        compounds (TPCs), 85, 102
        content (TPC), 41, 42, 44, 220
    starch content (TSC), 41
Traditional
    heat treatment, 219
    microwaving, 41, 42
    thermal pasteurization, 221
Transformer, 260–262, 269, 274, 275
Transient
    potential, 266, 275
    threshold, 215
*Trichinella spiralis*, 251
Trolox equivalent antioxidant capacity, 44
Turbulent velocity, 212
*Typhimurium*, 251
Typhoid, 117
Tyrosinase, 82

**U**

Ultra-high
    pressure (UHP), 226
    temperature (UHT), 1, 8, 12
Ultrasonic
    cavitation, 220
    waves (UW), 101, 178–181, 183, 192
Ultrasound, 54, 70, 101, 158, 159, 163, 166, 167, 171, 176, 178–185, 187, 190–192, 199, 200, 227, 245

assisted osmotic dehydration (UAOD),
    70, 96, 158, 159, 162–170, 176,
    180–183, 185–189, 191
  waves, 101, 159, 163, 171, 178, 182,
    185, 190
Ultraviolet (UV), 51, 85, 117, 127, 176,
  200
  radiation (UVR), 76, 117, 127

## V

*Vaccinium corymbosum*, 74
Vacuum drying, 76
Vanillin, 8
Vegetative bacteria, 49
Velocity, 74, 100, 149, 202–205, 208–210,
  212, 214, 221
Venture, 200, 201, 205, 208, 209, 222
Venturi, 200, 208, 209, 214, 215
*Vernonia amygdalina*, 95
Viscosity, 90, 132, 135, 137, 140, 149,
  151, 152, 215, 221, 234, 272
Volatile
  components, 84
  flavor compounds (VFCs), 85
  substances, 226
Volatilization, 47

Volumetric
  flows, 208
  heating, 27, 30, 37, 43

## W

Water
  activity (Aw), 1, 70, 71, 85, 158, 161,
    168, 170, 171, 186
  evaporation, 31, 33
  loss (WL), 177, 180, 182, 183, 185, 186,
    191, 192
  mobility, 31
  molecules, 28, 45, 47, 164, 180
  nonthermal sterilizer, 118
    electrical conductivity (EC), 119
    microorganism's destruction rate, 125
    total bacteria count, 120
    total coliform bacteria and E. coli,
      121
  vaporization, 72
Weight reduction (WR), 191
Wet-bulb temperature, 72
Winding direction, 262, 263, 270, 271, 275

## Y

Yeast, 219, 220, 231, 232, 250

Printed and bound by CPI Group (UK) Ltd, Croydon, CR0 4YY
23/10/2024
01777703-0001